Split Manufacturing of Integrated Circuits
for Hardware Security and Trust

Ranga Vemuri • Suyuan Chen

Split Manufacturing of Integrated Circuits for Hardware Security and Trust

Methods, Attacks and Defenses

 Springer

Ranga Vemuri
University of Cincinnati
Cincinnati, OH, USA

Suyuan Chen
University of Cincinnati
Presently with Apple Inc.
Cincinnati, OH, USA

ISBN 978-3-030-73447-3   ISBN 978-3-030-73445-9   (eBook)
https://doi.org/10.1007/978-3-030-73445-9

This Springer imprint is published by the registered company Springer Nature Switzerland AG
The registered company address is: Gewerbestrasse 11, 6330 Cham, Switzerland

To Vasanta, Bhargav, and Sarvani
for your love, support, and trust.
—*Ranga Vemuri*

To Xiaoxin, An'lin, and Su
for your love and support.
—*Suyuan Chen*

Preface

Globalization of the semiconductor supply chain over the past two decades resulted in several vulnerabilities which can be potentially exploited by attackers at various points in the chain to steal the integrated circuit (IC) design and cause harm to the interests of its rightful owners. Split manufacturing (SM) is the process of splitting an IC design and fabricating it at two or more foundries such that no single foundry has complete knowledge of the design. SM thwarts several potential attacks including reverse engineering and insertion of hardware trojans. Ever since SM was invented about 15 years ago, there has been a significant interest in the security benefits of SM from researchers, designers, and government agencies.

During the past decade, researchers have identified potential vulnerabilities in the process of splitting a design for SM and developed several attack scenarios to recover complete or partial design information from split designs. In addition, researchers have also developed methods for preparing split circuit designs to defend against these attacks. A comprehensive understanding of the principles behind these attacks and defense methods is necessary for designers who might consider SM to improve supply chain security.

This book is intended for graduate students, researchers, and designers who would like to understand the state of the art in SM methods, attacks, and defenses. We present in detail several SM attack and defense algorithms and design flows proposed during the past decade. For convenience, we have divided the attacks broadly into two categories: design constraint based and satisfiability based. Similarly, the defense methods are also divided into two categories: defenses against design constraint based attacks and defenses against satisfiability based attacks. While this classification helps in organizing our presentation, it is somewhat arbitrary in several instances. For example, some of the satisfiability based attacks also use design constraints and some of the defenses against satisfiability based attacks also effectively thwart design constraint based attacks.

We have organized this book into six chapters and an appendix as follows:

Chapter 1   In this chapter, we review typical IC design and fabrication flows and supply chain vulnerabilities. We introduce SM methods for 2D,

2.5D, and 3D ICs and discuss design flows for each. We discuss
the security benefits of SM. We introduce attack models intended to
discover the missing information from a split design.

Chapter 2   In this chapter, we introduce approaches for SM attacks and evaluation metrics. We discuss the following design constraint based
attacks: a proximity based attack, attacks based on extended proximities, a network flow model based attack, a machine learning
based attack, an attack using simulated annealing, a proximity based
mapping with net based pruning method, and a structural pattern
matching method. The first four attacks aim to reverse engineer the
complete design while the rest aim to insert malicious trojan circuits
at specific locations in the design.

Chapter 3   We begin this chapter with an introduction to defense metrics and
general defense methods. We then discuss the following defense
methods to secure designs against design constraint based attacks:
pin swapping, secure bipartitioning and placement, secure multiway
partitioning, placement perturbation, routing perturbation, concerted
wire lifting, netlist clustering, artificial routing blockage insertion,
and netlist randomization.

Chapter 4   In this chapter, we briefly review satisfiability (SAT) checking,
logic encryption, and the SAT attack against logic encryption.
We then discuss the following attacks: SAT based layout recognition, SAT attacks for reverse engineering of combinational and
sequential circuits, hybrid attacks combining SAT methods with
proximity information, and SMT (satisfiability modulo theories)
based improvements to reduce the complexity of the SAT attacks.
The first attack aims to insert a trojan circuit in the IC and the rest
aim to reverse engineer the complete design of the IC.

Chapter 5   In this chapter, we discuss several defense methods that secure IC
designs from satisfiability based attacks as well as design constraint
based attacks. Specifically, we discuss the following methods: SAT
based greedy wire lifting, simultaneous wire lifting and cell insertion, combined layout camouflaging and SM methods for 2D and
3D ICs, combined logic encryption and SM, and obfuscated built-in
self-authentication.

Chapter 6   In this chapter, we summarize the existing challenges and review the
emerging directions for further research. In particular, we discuss
the following directions: splitting at higher-levels of abstraction, SM
for analog and mixed-signal designs, SM with novel devices, new
attack models, exploration of manufacturability vs security tradeoff,
and utilizing advances in machine learning.

Appendix A   In this appendix, we summarize the data pertaining to the various
benchmark suites used by SM researchers to evaluate the proposed
attacks and defense methods. We provide references to the sources
of these benchmarks.

This book owes the information contained in it to the researchers who have developed the respective methods. We thank them for their dedicated contributions to this topic. We encourage the readers to consult the original publications cited in this book for further information on the methods and their experimental evaluations.

Cincinnati, OH, USA

February 22, 2021

Ranga Vemuri

Suyuan Chen

Acknowledgments

We are grateful for the encouragement and support we have received over the years from Mr. P. Len Orlando III, AFRL; Dr. Brian Dupaix, AFRL; Dr. Matt Casto, USAF; and Dr. Praveen Chawla, Edaptive Computing Inc., to foster our research in hardware security and trust.

We are thankful for the stimulating research environment at the University of Cincinnati, in particular for the interactions with Professors Marty Emmert, Rashmi Jha, Wen-Ben Jone, and Carla Purdy on various topics in hardware design and security.

We are fortunate to have worked with many dedicated graduate students on VLSI design and design automation. Related to the topic of this book, Rongrong Liu, Michael Shuster, Linden Peterson, Andi Thomas, and Daniel Mraz have worked in our lab on various problems in split manufacturing. We are thankful for the insights gained due to discussions with them. We thank Charles B. Glaser, Editorial Director; Olivia Ramya Chitranjan, Project Coordinator; Kala Palanisamy, Project Manager; and the entire Springer editorial and production teams for their excellent and timely help during the process of making this book a reality.

Contents

About the Authors

Ranga Vemuri has been on the faculty of the Electrical Engineering and Computer Science Department at the University of Cincinnati since 1989 and is currently a professor. He directs the Digital Design Environments Lab. His interests span various topics within hardware trust, correctness and security; VLSI design and architectures; formal methods and formal verification; electronic design automation; and reconfigurable computing and FPGAs. He and his students have published over 300 papers and have received several best paper awards and nominations. Prof. Vemuri guided 42 PhD and 90 MS students. His research has been funded by AFRL, DAGSI, DARPA, NSF, State of Ohio, and various industries including EDAptive Computing Inc. Prof. Vemuri was an associate editor of the *IEEE Transactions on VLSI* and a guest editor of the *IEEE Computer*.

Suyuan Chen received a PhD degree in electrical engineering from the University of Cincinnati in 2019. Currently, he is an ASIC engineer at Apple Inc., where his work deals with SoC design using state-of-the-art 7nm, 5nm, and 3nm CMOS technologies. His interests span a variety of topics in hardware security, LSI/AMS IC design methodologies, and computer aided design.

Acronyms

3PIP	3rd party intellectual property
AC	Attack correctness
ALD	Average logical difference
ALU	Arithmetic logic unit
AMSPR	Average mapped set pruning ratio
ANHD	Average normalized hamming distance
AOCS	Average output cone size
ASIC	Application-specific integrated circuit
AT	Arrival time
BB	Boundary box
BCP	Boolean constraints propagation
BEOL	Back-end-of-line
BGS	BEOL-driven gate selection
BISA	Built-in self-authentication
BIST	Built-in self-test
CA	Classification accuracy
CAD	Computer-aided design
CCR	Correct connection rate
CDCL	Conflict-driven clause learning
CI	Composition index
CMOS	Complementary metal-oxide silicon
CMP	Chemical-mechanical polishing
CNF	Conjunctive normal form
CNT	Carbon nanotube
CNTFET	Carbon nanotube field-effect transistor
CUT	Circuits under test
DAC	Digital-to-analog converter
DES	Data encryption standard
DfT	Design for trust
DFT	Design for testability
DIP	Distinguishing input pattern

DL	Deep learning
DNN	Deep neural network
DPLL	Davis–Putnam–Logemann–Loveland
DRC	Design rule check
EC	Elevating cells
ECO	Engineering change order
EDA	Electronic design automation
EEPROM	Electrically erasable programmable read-only memory
ELVT	Extreme low threshold voltage
EMSR	Effective mapped set ratio
EPL	Embeddable programmable logic
EPROM	Erasable programmable read-only memory
ERC	Electronic rule check
F2B	Face-to-back
F2F	Face-to-face
FEOL	Front-end-of-line
FF	Flip-flop
FGMOS	Floating-gate MOS
FI	Fault impact
FinFET	Fin field-effect transistor
FM	Fiduccia–Mattheyses netlist partitioning algorithm
FOM	Figure-of-merit
FPGA	Field-programmable gate array
GAAFET	Gate all-around field-effect transistor
GF	Global Foundries Inc.
GSHE	Giant spin-hall effect
HD	Hamming distance
HPWL	Half-perimeter wire-length
HT	Hardware trojan
IARPA	Intelligence advanced research projects activity
IC	Integrated circuit
ICR	Incorrect connection rate
ILP	Integer linear programming
IO	Input and output
IoT	Internet of things
IP	Intellectual property
LC	Layout camouflaging
LFSR	Linear feedback shift register
LGS	Logic-aware gate selection
LoC	List of candidates
LPP	Logic-driven placement perturbation
LR	Lagrangian relaxation
LUT	Look-up table
LVS	Layout vs schematic
LVT	Low threshold voltage

MgO	Magnesium oxide
MI	Mutual information
MILP	Mixed integer linear programming
ML	Machine learning
MUX	Multiplexer
NY	Neighbor connectivity
NC	No cycle
NGSLD	Neighbor gates with significant logical difference
NHD	Normalized hamming distance
NoC	Network-on-chip
NOCS	Normalized output cone size
NSEO	Normalized signal effect on output
OBISA	Obfuscated BISA
OER	Output error rate
ORA	Output response analysis
PC	Placement congestion
PDK	Process design kit
PNR	Percentage of netlist recovered
POC	Primary output cone
POS	Product of sum
PPA	Power, performance, and area
PPP	Physical-driven placement perturbation
RAT	Required arrival time
RC	Routing congestion
RDL	Redistribution layers
RF	Radio frequency
RRAM	Resistive random-access memory
RO	Ring oscillator
RTL	Register transfer level
SA	Simulated annealing
SADP	Self-aligned double patterning
SAT	Satisfiability
SCOAP	Sandia controllability/observability analysis program
SEO	Signal effect on output
SES	Statistical element selection
SM	Split manufacturing
SMT	Satisfiability modulo theories
SoC	System-on-chip
SRAM	Static random-access memory
SSI	Stacked silicon interconnect
SVT	Standard threshold voltage
TIC	Trusted integrated chips
TPG	Test pattern generator
TSV	Through silicon via
ULVT	Ultra low threshold voltage

VeSFET	Vertical slit field effect transistor
vpin	Virtual pin
VLSI	Very large scale integration
WLD	Weighted logical difference

List of Figures

List of Tables

Chapter 1
Split Manufacturing Methods

Abstract Modern integrated circuit (IC) design and fabrication processes are both complex and expensive. In order to reduce the cost and improve efficiency, supply chains based on fabless design houses and pure-play foundries have evolved. Globalization of IC supply chains introduced several *vulnerabilities* which can be exploited by unscrupulous attackers. Split manufacturing (SM), originally introduced for yield enhancement, offers a defense to protect against some of these vulnerabilities. This chapter reviews typical IC design and fabrication processes and discusses various vulnerabilities in the globalized supply chains. We introduce 2D, 2.5D, and 3D split fabrication methods, their potential security benefits, and the design flows for each method. We discuss the basic issues and tradeoffs in SM and summarize several SM hardware demonstrations. We introduce potential attack scenarios against SM and discuss the objectives of the attackers. We classify the attacks based on the methods used and describe the assumptions made. The remaining chapters in this book discuss the proposed attacks and defense methods in detail.

1.1 Integrated Circuit Design Process

A typical integrated circuit (IC) design process is illustrated in Fig. 1.1. While many variations of this process exist, this prototypical process is sufficient for our purposes.

IC design process usually begins with a register-transfer level (RTL) model in a hardware description language such as Verilog. Often, a variety of pre-designed Intellectual Property (IP) cores in the form of synthesizable RTL modules are integrated into the design at this stage. These IP cores are termed *soft* cores. Following satisfactory simulation results, a logic synthesis tool is used to produce a gate-level design subject to power, performance, and area (PPA) constraints. A library of gate-level components is used during the synthesis process. These components are pre-characterized for performance and power under various output load and input slew conditions. These characterization data are used during the technology mapping step of logic synthesis to bind the abstract logic gates to

Fig. 1.1 Integrated circuit design process

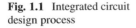
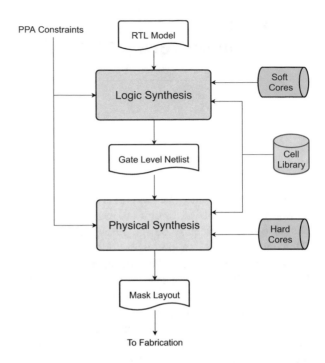

concrete cells available in the library. The synthesized gate-level design, usually expressed as a structural model in Verilog, is simulated to ensure that it is functionally correct and to characterize its performance and power before layout. In addition, timing and power analysis tools can be used to determine performance and power characteristics of the design using the library data.

Following this step, a physical design synthesis tool is used to place and route the cells in the synthesized netlist to produce a mask layout which should conform to the layout design rules specified by the would be manufacturer of the IC. The layout synthesis process usually consists of two steps: placement and routing. During the placement process, pre-existing library cells, termed *standard cells*, are placed in a rectangular area representing the surface of the die. Some modules, termed *hard* IP cores, may pre-exist and be integrated with the rest of the layout using macro placement and routing methods. During routing, wires to connect the standard cells, and the hard macros, if any, are generated per the netlist connectivity requirements. The entire layout generation process is driven by performance considerations and input and output (IO) terminal connectivity requirements. After a layout is generated, a transistor-level netlist, including various parasitic capacitor, resistor and, possibly, inductor components, is extracted and simulated using a circuit simulator to ensure continued satisfaction of performance and power requirements.

Over the past three to four decades, *CMOS* (Complementary Metal-Oxide Silicon) has been the dominant solid-state medium for implementing electronic circuits. Although many topics discussed in this book are independent of the implementation technology, we will discuss these topics in terms of a prototypical CMOS process. Digital CMOS circuits consist of two types of transistors, the n-type and the p-type, and wires to connect them together. Accordingly, in ICs targeted for implementation in a CMOS process, layouts consist of two types of layers: *device* and *interconnect*. Device layers are comprised mainly of well, diffusion, and poly-silicon layers. Wires in these layers together define the sizes and locations of the n-type and p-type transistors in the design. Interconnect layers carry the metal wires for connecting the devices. Layout of rectangles in these layers denotes the shape and location of the wire fragments and contacts among these wire fragments. Contacts among the interconnect layers are called *vias*. While older processes allowed only two or three metal layers for interconnect, state-of-the-art CMOS processes allow over a dozen metal layers.

For more details about the IC design process, the readers may refer to standard text books on CMOS IC design [1–3] and electronic design automation (EDA) [4–7].

1.2   Integrated Circuit Fabrication

Once the mask layout is ready, the artwork is sent in the form of a GDSII file to a fabrication (fab) facility, also called a *foundry*, where the IC is manufactured. Alternatively, a standard cell netlist can be sent to the manufacturer and the layout can be produced at the fab. We discuss the steps in a typical CMOS fabrication process illustrated in Fig. 1.2.

IC fabrication begins with the preparation of slices, called *wafers*, of monocrystalline semiconductor material. The fabrication process involves repetitive application of three types of operations to build a circuit on the wafer: *doping, deposition,* and *etching*. Doping is the process of implanting n-type or p-type dopant materials in selected regions of the wafer where transistors need to be formed. Deposition is the process of depositing layers of conducting or insulating material across the wafer. Etching is the process of removing material. Chemical etching is used to

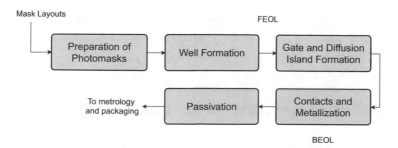

Fig. 1.2   Integrated circuit fabrication process

remove material selectively according to the layout geometry. Chemical etching or chemical-mechanical polishing (CMP) can be used to remove material across the entire wafer surface.

These operations are guided by the layout geometry sent to the fabricator from the design house. In order to guide selective application of these steps, a process called *photo-lithography* is used. In this process, first, the rectangular geometrical shapes defining various layers in the layout are transferred onto a set of glass sheets called *photo-masks* or, simply, *masks*, or *reticles*. The wafer is coated with a light sensitive film called *photo-resist* or, simply, *resist*. When a mask is exposed to light, the image of the geometrical shapes patterned on the mask is projected on the resist. When positive (negative) resist is used, exposed regions are removed (remain) and unexposed regions remain (are removed) after development. The remaining resist is used as the mask for the next step. Optical alignment methods are used to precisely align various masks to the wafer in each step so as to correctly and precisely transfer the layout onto the wafer.

The fabrication process can be divided into two distinct phases. In the first phase, termed *front-end-of-line* (FEOL), transistors and other passive devices, if any, are implemented at the required locations using selective oxidation and doping steps. In the second phase, termed *back-end-of-line* (BEOL), these devices are connected using several metal layers interspersed with oxide layers and vias between the layers. A cross-sectional view of a typical interconnect structure is shown in Fig. 1.3. For each metal layer in the interconnect, two physical layers are needed. An oxide layer is used to separate neighboring metal layers. Vias are patterned at the required locations in the oxide layer. Required metal wires are then patterned on top of the oxide layer. Entire routing structure in the layout is implemented in this manner. Table 1.1 shows pitch values (minimum wire width + minimum spacing between neighboring wires) for various metal layers in the Intel 10 nm process [8].

Using these steps, hundreds of ICs arranged in several rows are simultaneously produced on each wafer. After processing, the wafer is cut into individual *chips* which are then packaged. Following functional and performance evaluation, much of which can occur before packaging, the chips are ready for use. Chips may be sold either in the packaged form for system integration on a printed circuit board or in the unpackaged *bare die* form for system integration on other forms of multi-chip carriers [10].

While our highly simplified description is sufficient to follow the topics in this book, interested readers may refer to the excellent texts on IC fabrication for details [11–14].

1.3 Globalization of the IC Supply Chains

Capacity and complexity of integrated circuits have been steadily increasing since the advent of CMOS in early 1970s. The famed Moore's Law [15] had correctly predicted a doubling of the number of transistors per die every 2 years

Fig. 1.3 Cross-sectional view of typical interconnect layers [9]

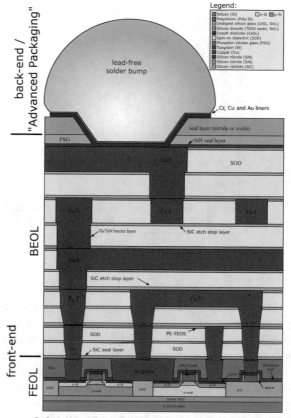

By Cepheiden - self made (from university scripts and scientific papers), CC BY 2.5, https://commons.wikimedia.org/w/index.php?curid=1445444

Table 1.1 Metal pitch values for Intel 10 nm process [8]

Metal layer	M0	M1	M2	M3	M4	M5	M6	M7	M8	M9	M10
Pitch (nm)	40	36	44	44	44	52	84	112	112	160	160

or so, attained primarily by reducing the smallest manufacturable feature size by approximately half. Figure 1.4 shows the transistor count and feature size trends [16]. Today, over 50B transistors can be incorporated on a single die in a typical 7 nm CMOS technology node. 5 nm products are expected shortly and semiconductor manufacturers have already announced plans for commercial deployment of 3 nm and 2 nm process nodes. While geometrical scaling of planar devices has been the trend until early 2000s, introduction of new materials and newer "vertical" structures such as FinFETs and GAAFETs [17, 18] have helped maintain Moore-*equivalent* scaling since then. Full *3D integration*, discussed later in this chapter, is expected to continue to ensure the life of the CMOS as a viable medium for integrated circuits for the next 10–15 years.

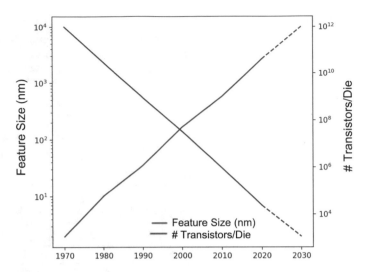

Fig. 1.4 Exponential trends in feature size reduction and transistors/die growth

The growing capacity and complexity of CMOS ICs resulted in two important *globalization* trends: design *reuse* and *fabless* design companies. Since designs have become extremely complex, in order to keep the design time and design cost under control, design methodologies and processes that encourage reuse of known-good modules across all levels of design abstraction have been developed. The most important of these practices is the use of soft or hard cores developed by other companies. Driven by a globalized economy, these so-called *3rd party IP cores* (3PIP) are often developed by companies incorporated or physically located in other countries. It is not uncommon for a system developer to procure 3PIP cores from multiple foreign countries. Mishra et al. [19] show that the geographical distribution of IP cores integrated in representative IC products could span across multiple continents. Even when multiple cores developed within the same company are brought together in a system, those cores may have been developed by multiple development teams located in multiple countries by employees many of whom could be foreign nationals.

IC fabrication facilities for advanced CMOS nodes are exceedingly expensive, costing upwards of $10B for establishing a new state-of-the-art production plant with additional recurrent operating costs. This resulted in a situation where it is no longer cost-effective for most IC design companies to own and maintain a foundry exclusively devoted to their own products. On the other hand, to remain profitable, a fabrication facility needed to produce sufficient quantities of ICs possibly for multiple design houses so that the fab can be kept busy. These considerations led to *fabless* IC design companies that develop state-of-the-art IC products but do

Table 1.2 Top 15 semiconductor sales leaders in 2020 [20]

Rank	Company	Type	HQ
1	Intel	IDM	U.S.
2	Samsung	IDM	South Korea
3	TSMC	Foundry	Taiwan
4	SK Hynix	IDM	South Korea
5	Micron	IDM	U.S.
6	Qualcomm	Fabless	U.S.
7	Broadcom Inc.	Fabless	U.S.
8	Nvidia	Fabless	U.S.
9	TI	IDM	U.S.
10	Infineon	IDM	Europe
11	MediaTek	Fabless	Taiwan
12	Kioxia	IDM	Japan
13	Apple	Fabless	U.S.
14	ST	IDM	Europe
15	AMD	Fabless	U.S.

IDM integrated device manufacturer

Table 1.3 Top fabless IC design companies in 2Q20 [21]

Rank	Company	HQ
1	Broadcom	U.S.
2	Qualcomm	U.S.
3	Nvidia	U.S.
4	MediaTek	Taiwan
5	AMD	U.S.
6	Xilinx	U.S.
7	Marvell	U.S.
8	Novatek	Taiwan
9	Realtek	Taiwan
10	Dialog	U.K.

not own a fab and *pure-play* foundries that do not produce significant IC products of their own but offer their fab services to other, usually fabless, design houses. Tables 1.2, 1.3, and 1.4 show data pertaining to the top semiconductor sales leaders, fabless design houses, and pure-play foundries. Three trends are evident from this data: (1) only a few leading IC design companies own in-house fab facilities, (2) many fabless design houses are among the top IC developers and several are headquartered in the USA, and (3) all, except one, of the pure-play foundries are located in countries other than the USA. Taken together these data show the geographical dispersion of the *IC supply chains*. It is worth noting many ICs used in military systems as well as consumer products are produced by the same supply chains.

Table 1.4 Top pure-play
foundries in 2Q2019 [22]

Rank	Company	HQ
1	TSMC	Taiwan
2	Samsung (IDM)	South Korea
3	GlobalFoundries	U.S.
4	UMC	Taiwan
5	SMIC	China
6	TowerJazz	Isreal
7	Hua Hong	China
8	VIS	Taiwan
9	PSC	Taiwan
10	DB HiTek	South Korea

1.4   Vulnerabilities in the IC Supply Chains

The globalization trends discussed in the previous section have no doubt fueled the phenomenal growth of IC products while reducing their cost and fostered unprecedented and disruptive advances in multiple application areas ranging from consumer products such as mobile devices, game systems, and Internet-of-Things (IoT) gadgets to high-performance applications such as data centers, cloud computing, and machine learning. However, the same globalization trends led to potential vulnerabilities related to security and *trust*. Vulnerabilities, when exploited by malicious attackers, cause harm to the legitimate interests of the rightful owners of the design, intellectual property, product, or data.

Broadly speaking, *security* is concerned with protection from unauthorized access to and use of privileged data or intellectual property. *Trust* is concerned with the confidence that a process or product works as expected and does not take unfair advantage of its users. A global IC supply chain admits or exacerbates several vulnerabilities which make it potentially untrusted and unsecured leading to various attack scenarios as shown in Fig. 1.5. Some of these threats are listed below:

1. **Trojan Insertion:** A *hardware trojan* (HT) is a malicious addition or modification to the design done with the intent of crashing the system during operation, producing incorrect results, or stealing privileged data. While trojans causing denial of service can be inserted by mere access to the layout or netlist, to insert data-stealing trojans an attacker would require detailed knowledge of how the design works. Trojans can be inserted almost at any point in the supply chain until fabrication commences [23–25].
2. **IP Piracy:** Designs can be stolen at various levels of abstraction through the supply chain. User of an IP or a rogue actor at a foundry may pirate the IP and gain unfair advantage. The pirated IP can be used to produce illegal products identical to the originals or improved with modifications or infected with malicious spyware [19, 26].

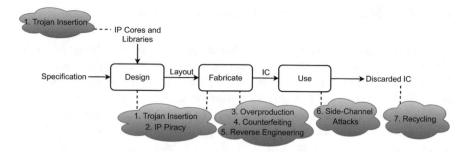

Fig. 1.5 IC supply chain threats

3. **Overproduction:** An untrusted foundry may produce more ICs than the quantity ordered by the design house and sell the excess production through illegal market channels. Such unauthorized production and use obviously comes at the expense of the product's rightful owner's profit margin.
4. **Counterfeiting:** Stolen or reverse engineered IP can be used to produce counterfeit ICs, often inferior in reliability and performance to the originals. These counterfeit ICs can be channeled into the supply chain leading to cheap but untrustworthy products [27–29].
5. **Reverse Engineering:** Reverse engineering is the process of starting with the physical IC and determining the layout by depackaging, delayering, and imaging each layer, and stitching the images together to produce the mask layout of the IC [30, 31]. Once the layout is reconstructed, the attacker may proceed to extract a netlist and reverse engineer the design to a desired level of abstraction. Since the foundry has access to the entire layout to begin with, an unscrupulous actor at an untrusted foundry may reverse engineer the design from the layout and use it in various ways. Note that delayering is an expensive, tedious, and destructive process. In addition, use of novel devices [32] and layout *camouflaging* methods [33–37] can make effective delaying quite difficult.
6. **Side-Channel Attacks:** Although not directly related to the globalization of the supply chain, side-channel attacks have emerged as a threat especially against small embedded systems. In this form of attack, an attacker measures and uses power, radiation, or other forms of non-functional output parameters of the IC to extract critical data during run-time. Side-channel attacks are mounted directly on the end-product after IC production, system integration, and product sale [38–40].
7. **Recycling:** ICs can be scavenged from older, possibly discarded, systems and recycled into the market. Such recycled ICs may not be functional or may be at the end of their reliable life. When such ICs are procured and used in safety-critical systems, they may cause serious harm upon failure [41, 42].

Interested readers may refer to [43, 44] for additional details on these and other hardware security and trust issues in the IC supply chains and methods of mitigation.

1.5 Split Manufacturing of Integrated Circuits

Split manufacturing (SM) of ICs was introduced in early 2000s by Jarvis and McIntyre as a means for yield enhancement, disclosed in a patent assigned to Advanced Micro Devices Inc. [45, 46]. *Yield* is the ratio of the number of functional dies to the total number of dies of an IC manufactured in a wafer run. The motivation behind SM was to separately fabricate the FEOL and BEOL parts, test the two parts separately, and then merge the known-good parts into a finished IC through an alignment and *bonding* procedure. In this way, parts with faults in the BEOL metal layers are eliminated and only fault-free BEOL parts are bonded to the parametrically tested good pieces from the FEOL process leading to a net increase in yield.

Various instances of SM were postulated based on this general idea. For example, the BEOL and FEOL layers could follow different design rules. They can even be manufactured in different foundries. BEOL metal wires and vias could be made larger so as to enhance their current carrying capacity and reliability. In addition to the devices, the FEOL part may consist of one or more metal layers and the rest of the metal layers could be delegated to the BEOL part. It is also possible to connect one FEOL piece with multiple BEOL pieces or even multiple FEOL pieces with multiple BEOL pieces. It is recommended that the FEOL part should have at least one metal layer since it is then possible to form intra-cell interconnections among the transistor devices which allows functional testing of the gates and reliable formation of connections between the top metal layer in the FEOL part and the bottom metal layer in the BEOL part.

Clearly, the design and fabrication process for SM will have to be different than the one used for the traditional process. The process for SM is illustrated in Fig. 1.6. At some point before or during the physical design stage, the FEOL and BEOL designs should be separated. This separation is achieved through the identification and rerouting of some wires through the BEOL layers. This process is called *wire lifting* or *net lifting*. The FEOL part consists of all the transistors. When at least one metal layer is allowed in the FEOL part, as is usually the case in SM, all the logic gates are in the FEOL part. It may even consist of some nets or portions of nets depending upon the number of metal layers allowed in the FEOL part. The remaining nets or portions of nets are to be routed through the metal layers in BEOL part. Once separated, layouts of both the parts are generated according to the respective design rules. The FEOL part consists of the device layers and one or more of the lower level metal layers. The BEOL part consists of the remaining upper level metal layers. Extracted netlists from both parts can be connected together to form a complete circuit for analysis purposes. Following satisfactory simulations, layouts of the two parts are sent to the respective fab lines. When traditional planar or 2D technology is used, the FEOL part is first fabricated at the advanced open foundry and the BEOL fabrication is completed at the secure foundry using a *compatible* process. When using 2.5D or 3D integration, both parts are manufactured and independently tested, faulty parts are discarded, and good FEOL parts are combined

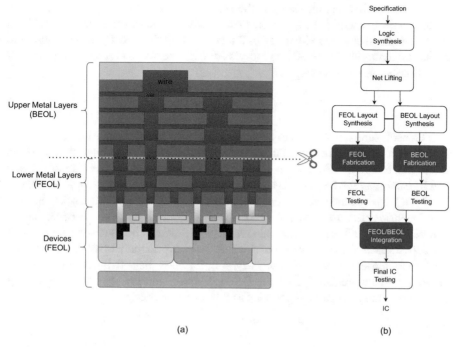

Fig. 1.6 Split manufacturing for 2D ICs. (**a**) 2D IC interconnect layers, (**b**) SM process for 2D ICs

Fig. 1.7 SM compatibility between two foundries (based on [47])

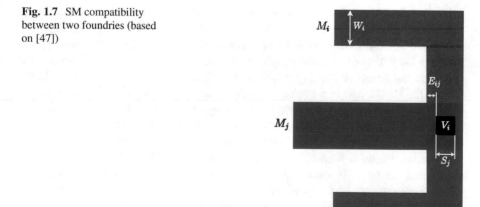

with good BEOL parts to form complete ICs. Following functional testing and packaging the ICs are ready for use.

Vaidyanathan et al. [47] introduced a notion of compatibility between two foundries for SM based on the alignment requirement between the BEOL and FEOL parts. Referring to Fig. 1.7, let M_i be the last (top) metal layer in the FEOL foundry and M_j be the first (bottom) layer in the BEOL foundry. Usually $j = i + 1$. Let W_i be the minimum width of M_i per the design rules of the FEOL foundry. Let S_j be the minimum size of the via of M_j in the BEOL foundry and E_{ij} be the minimum enclosure required of the FEOL foundry's M_i over the via of M_j. The two foundries are *SM compatible* provided,

$$W_i \geq S_j + 2 * E_{ij} \tag{1.1}$$

The enclosure size E_{ij} has to be agreed upon by the two foundries based on the lithographic misalignment tolerances between M_i in the FEOL foundry and M_j in the BEOL foundry. If Eq. 1.1 is not satisfied, then alignment between the BEOL and FEOL parts may not be possible and the two foundries are deemed incompatible.

If Eq. 1.1 is satisfied and the two foundries have identical BEOL stacks, then the two foundries are *fully compatible*. Any hard 3PIP cores that use the BEOL layers can be allowed and the same physical synthesis flow can be used for both parts. If Eq. 1.1 is satisfied but the two foundries do not have identical BEOL stacks, then the two foundries are *partially compatible*. In this case, hard IP blocks using BEOL layers need to be requalified for the BEOL foundry and the physical design flows have to be adapted separately for the FEOL and BEOL designs.

1.6 Security Benefits of SM

While originally proposed for yield enhancement, potential security benefits of SM were noticed soon after its introduction, for example, within the Intelligence Advanced Research Projects Activity's (IARPA) Trusted Integrated Chips (TIC) program [48]. The idea is quite simple: Manufacture the FEOL part at a potentially untrusted but advanced foundry which may be located in a foreign country. Manufacture the BEOL part in a trusted foundry located within the country but possibly supporting an older process node which may be sufficient for the performance requirements expected of the nets assigned to the upper level metal layers. Then, combine both the parts at the trusted foundry. Although, a malicious actor at the potentially untrusted foundry who has access to the FEOL layout can reverse engineer a netlist, it is incomplete due to the missing BEOL interconnect, and, hence, it cannot be used for unfair advantage or with malicious intent. Let us revisit each of the supply chain risks discussed in Sect. 1.4 in light of SM:

1. **Trojan Insertion:** Since the attacker located at the foundry does not have access to the full netlist, it is hard to gain sufficient design knowledge to insert an intelligent data-stealing trojan. However, it is still possible to insert trojans to deny service. In addition, SM does not prevent attackers located at the design house or the IP development sites from inserting trojans.
2. **IP Piracy:** Neither the entire design nor significant parts of the design, such as the IP cores, can be pirated at the FEOL foundry since complete netlist

connectivity information is unavailable. In fact, the net lifting process should ensure that any significant module in the design cannot be compromised.

3. **Overproduction:** Although the untrusted foundry may still *overbuild* the FEOL parts, it cannot build complete functional ICs due to the missing BEOL signals.

4. **Counterfeiting:** Since complete design or significant IP cannot be stolen at the FEOL foundry, counterfeiting resulting in unreliable and poor quality ICs is not possible.

5. **Reverse Engineering:** Since the FEOL layout has missing BEOL nets, an attacker having access only to the FEOL part or layout cannot reverse engineer a complete netlist by physical delayering and imaging or by extraction from the layout. Destructive *reverse engineering* of the completed ICs is possible if the attackers can obtain them but it is an expensive and error-prone process.

6. **Side-Channel Attacks:** SM does not prevent side-channel attacks. If attackers have access to complete ICs and systems in which the ICs are deployed, then side-channel attacks can occur.

7. **Recycling:** SM does not address recycling. If scavengers have access to discarded ICs and products in which the ICs are used, recycling can occur.

1.7   SoC Design Methodology for SM

System-on-Chip (SoC) designs integrate IP cores obtained from multiple sources. Two types of IP cores are prevalent: *soft* and *hard*.

Soft IP cores are provided in the form of synthesizable RTL models in a hardware description language such as Verilog or VHDL along with constraints and configuration scripts. Soft cores are highly parameterized in terms of various signal vector sizes and functions performed by the cores. This makes them useful in a wide variety of application contexts. In addition, since complete RTL is available, synthesis tools are free to cross-optimize the designs across core boundaries and various test generation and formal verification tools can be used for effective analysis.

Hard cores are provided in the form of pre-existing layouts. While a HDL simulation model may be provided, it may not be synthesizable and may not have any gate-level correspondence to the implementation. Thus, hard cores are "black-box" modules which bypass the logic synthesis step and are directly dropped into the global place and route stage in physical design. Hard cores would have already been validated for design rules and functional correctness and characterized for performance and power. In some cases, hard cores may have already been demonstrated through fabrication and post-silicon validation. Hence, hard cores save significant design time and cost while increasing the chances of the SoC operating correctly after first-time integration and fabrication. Due to this reduction in silicon bugs and product-to-market time, hard cores emerged to be a popular choice for SoC integrators.

EDA (Electronic Design Automation) tools for SoC designs targeted to a single foundry use the Process Design Kit (PDK) of that foundry for design rules, process and transistor models, parasitic extraction data, etc. Use of soft cores in SoC designs targeted for SM at two foundries, both using the same process, is straightforward. The design process can continue to use the PDK of that process. However, when two different processes at two different foundries are used, the following modifications are necessary to the design flow: In the logic synthesis step, libraries from the FEOL foundry should be used. Since all devices are fabricated at the FEOL foundry, schematic design, if any done, should be based on the PDK from the FEOL foundry. In the physical design step, LVS (Layout vs Schematic Verification) and DRC (Design Rule Checking) should pass with the PDK from the FEOL foundry. In addition, after all the BEOL layers are identified, the LVS and DRC corresponding to those layers should pass with the PDK of the BEOL foundry. As an additional step, wherever the top FEOL metal wires are connected to the bottom BEOL layer, compatibility rules between the two foundries should be verified through appropriate cross-foundry DRC.

When hard cores are used, the IP core vendor needs to provide the necessary support for SM. This requires the IP vendor to work closely with the SoC designers to determine the metal layer at which the split occurs and determine various design flow details accordingly. In particular, all FEOL layout should be produced and validated using the PDK from the FEOL foundry and the BEOL layout should similarly conform to the PDK of the BEOL foundry. It is assumed that the two foundries are SM compatible as discussed earlier.

1.8   Where to Split? Which Wires to Lift?

The basic question in SM is where to split? The IC is said to be *split at metal level* M_i if M_i is the first (i.e. bottom) layer in the BEOL part. All the layers below M_i are assumed to be in the FEOL part. A closely related question is, which wires to lift? Usually, lower levels of metal are used for intra-cell routing and the upper levels of metal are used for inter-cell routing. In general, splitting at a lower level implies higher security but at the expense of fabrication cost and performance. Conversely, splitting at a higher level leads to lower security since relatively few wires are hidden from the attacker but BEOL cost is reduced and there is little to no performance penalty.

For example, splitting at M_2 implies essentially all of the inter-cell wires are lifted into the BEOL. These wires include critical paths. While ensuring a high degree of security, this has serious consequences on the design performance. In addition, lower levels of metal layers have smaller pitch and carry higher routing density which increases the fabrication cost. This requirement contradicts the basic motivation to use SM for security since it is assumed that the secure BEOL foundry is cost-effective since it needs to support only an older process node.

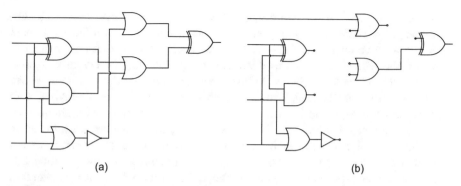

Fig. 1.8 Wire lifting. (**a**) Original circuit, (**b**) Circuit after wire lifting

Vaidyanathan et al. [47] proposed a simple metric to quantify the potential benefit of lifting some n wires in the BEOL part. Assume that each lifted wire is a two terminal connection with one driver and one receiver. For each lifted wire then a driver end, say, the output of a gate, and a receiver end, say, the input of a gate are left unconnected in the FEOL part, as illustrated in Fig. 1.8. Lifting of n wires results in n driver ends and n receiver ends without connections. Assume that the attacker, who has access to the FEOL netlist, can identify these *hanging* terminals in the netlist. There are $n!$ ways to connect the n driver (output) terminals to the n receiver terminals. In the absence of any other information, the attacker is confronted with a daunting number of options only one of which is correct. Even in the presence of equivalent subcircuits, relatively few of the possible connection patterns will be correct. In addition, for an attacker who is not in possession of a working chip or correct model of the design, determining *whether* the correct design is recovered is difficult. Even a relatively small n should quickly reduce the odds of a successful attack. However, as we will discuss later, due to the development of clever attack methods, the specific wires lifted must be selected with care. How many wires to lift, exactly which wires should be lifted, what is the security guarantee, and what are the performance, power, or other penalties? These are the key questions in any wire lifting methodology for SM.

1.9 Hardware Demonstrations of Secure SM

Soon after the potential security benefits of SM were recognized, several researchers demonstrated split fabricated IC designs. These demonstrations which appeared during 2013–2016 are summarized below.

1. *Global Foundries—IBM 130 nm SM Demonstrations:*
   Vaidyanathan et al. [47, 49] demonstrated SM using two semi-compatible 130 nm processes at two different foundries: Global Foundries (GF), Singapore, was used

as the "untrusted" FEOL foundry and IBM, Burlington, as the "trusted" BEOL foundry. The device and M1 layers were fabricated at GF and all the subsequent layers were completed at IBM. They have developed a custom design flow in which the schematic entry was done in the GF PDK and the layout was done such that the layers at or below M1 used the GF PDK and the layers above M1 used a superset of the design rules of both PDKs so that the same layout, with minimal changes, could be used to compare the SM designs with single flow designs fabricated entirely at GF. Based on this, they have also developed an ASIC/SoC design flow in which the physical synthesis and extraction/verification steps were adapted for SM. The M1 layer was precluded for use for inter-cell routing. Empty space after layout synthesis was filled with (unused) standard cells so that the attacker cannot distinguish between the design cells and the filler cells and cannot insert a trojan due to lack of space.

To demonstrate secure SM with split after M1, Vaidyanathan et al. [47] designed and fabricated ring oscillator (RO) test chips using the SM flow and the single flow (only GF). A 32-bit multiplier was synthesized and configured as a RO. Both runs yielded fully functional chips with negligible differences between two test chips. RO frequencies measured at various supply voltages were almost the same, well within the small differences allowed by wafer-to-wafer process variations.

In addition, Vaidyanathan et al. [47] implemented an imaging test chip containing a sub-block used in high-performance imaging applications. This was designed with standard cells and SM-compatible SRAM IP blocks (discussed below). Again, measurements of power dissipation at various supply voltages showed no significant differences between the SM flow chips and full flow chips. Similarly, there was no significant performance *overhead* for SM for these foundries for these designs (number of gates ranging from 1624 to 5673). Since all inter-cell nets were lifted, for n gates, there are $n!$ possible connections. This poses a daunting challenge to an attacker in the absence of further exploitable information and guarantees a high level of security.

Vaidyanathan et al. [50] also proposed a method of using sacrificial test-only BEOL stacks to exhaustively test the FEOL dies and detect reliability attacks designed to trigger a trojan during IC operation. Simulation results showed that the approach ensures that the tested FEOL dies were functionally correct and free of trojans and the remaining FEOL dies can be integrated with the real BEOL stacks. They mentioned plans to validate the method through fabrication of chips using the 65 nm process at GF for FEOL including M1 and a similar 65 nm process at IBM for the remaining BEOL.

2. *Other Processes and Foundries:*
   Similar results, for much larger digital and analog blocks, were reported by the IARPA's TIC program [48, 51] when using the 65 nm nodes of GF and IBM foundries as well as two 28 nm nodes of Samsung (Korea) and Samsung (Austin). These demonstrations indicate that the security premise of SM does not necessarily involve large power or performance overheads and design retooling costs at least when two largely compatible processes are used. Further, these

demonstrations show the viability of splitting right after M1 when both processes are largely compatible and the fabrication cost of all the remaining metal layers at the BEOL foundry can be justified.

3. *SM for Digital and Analog IP Blocks:*
As a part of the above demonstrations, Vaidyanathan et al. [49] developed an SM design flow for IP blocks. Certain hard IP blocks have a standard structure which can be recognized and exploited by attackers. For example, in SRAMs, peripheral blocks such as bitline sensors and wordline drivers have predictable locations relative to the bitcell arrays.

Vaidyanathan et al. developed a method to synthesize obfuscated smart memories without using such predictable structures. The method uses (1) randomized placement of SRAM peripheral circuit cells, (2) minimizing or avoiding common topologies for peripheral circuits and using standard cells instead of the usual recognizable leaf cells, and (3) adding non-standard, application-specific functions to the SRAM to improve both performance and obfuscation. They have synthesized and fabricated a parallel 2×2 access 1 KB SRAM through the GF-IBM split 130 nm process as well as through the single flow GF process. Comparison between the two showed that both achieved similar peak frequencies. However, the SM version consumed 25% less area and 12% less power due to its application-specific circuitry which had more than offset any overheads due to SM.

Vaidyanathan et al. have also demonstrated secure analog IP using SM using three techniques during design: (1) adding dummy transistors in empty spaces for camouflaging the real transistors until M1, (2) regularizing transistor widths such that transistors with different lengths can abut each other so as to obscure their boundaries, and, (3) employing block routing between transistors below M1 to make it difficult to infer connections between transistors. In addition, the Statistical Element Selection (SES) [52] analog architecture design style, which employs transistor-level redundancies to allow post-fab tuning of performance metrics via an on-chip digital controller, was adopted for the design of the obfuscated analog IP. They have designed and fabricated a 14-bit SES current steering digital-to-analog converter (DAC) circuit in the same GF-IBM SM process as well as the GF single process. Again, they have reported minimal differences, entirely within the measurement noise, when the SM chips were compared with single flow chips.

4. *SM of an Asynchronous FPGA*
Hill et al. [53] demonstrated SM of an asynchronous FPGA. They have designed and fabricated a 1.3 million transistor, island-style, quasi delay-insensitive FPGA using both a split foundry process (foundry A for FEOL and B for BEOL) and a single foundry process (foundry A). While, both were 130 nm processes, foundry B offered 10–15% worse RC characteristics per unit area for thin wiring used for local interconnect and 5% better characteristics for think wiring used for long distance interconnect. The union of the most restrictive design rules from both foundries was used to make it easier to implement the design in both SM and single process flows. The 9 mm^2 FPGA die contained 5x5 array of

tiles where each tile contained 52,000 transistors. Each tile contained a logic block with four 4-input look-up tables, source and sink units for data tokens, and a programmable switch-box to connect the inter-tile channels in a 2D mesh network. 32 channels/tile in each direction were used within the pipelined routing fabric.

Hill et al. reported that all dies from both the SM and single process flows were functional. Static power consumption, mostly dependent on the FEOL devices, was almost the same for the dies from both processes. Dynamic power and performance, measured with the help of on-chip test configurations, were impacted by BEOL connections. Chips from Foundry A operated at an average frequency of 342 MHz consuming 20.3 pJ/operation while those from SM operated at 311 MHz consuming 21.2 pJ/operation. Overall, the performance cost of SM was about 10% and energy cost was about 5% compared to single foundry production. Note that neither flow made the best use of the foundries, since the design rules used were a conservative union of the rules of both foundries.

After these early and successful demonstrations, the attention of the research community shifted to the development of attacks against SM and effective defenses to thwart those attacks.

1.10   SM for 2.5D Integrated Circuits

In 2.5D integration [54], multiple bare dies, sometimes called *chiplets* are independently fabricated and are then interconnected together using a silicon interposer which consists of a signal *redistribution layer* (RDL), as illustrated in Fig. 1.9a. Dies are attached to the pads on one side of the interposer through the IO buffers in each die and the micro-bumps between the dies and the interposer. These pads are directly connected to *through silicon vias* (TSV) which pass through the interposer substrate and connect to the package substrate via flip-chip bumps.

2.5D integration offers several benefits over traditional 2D ICs. Each chiplet can be small resulting in high yield and reliability. Since each chiplet is independently fabricated, different technologies and process nodes can be used. For example, critical cores/modules can be implemented using a state-of-the-art process node and others can be pre-fabricated dies from an older process node. In contrast, 2D integration forces the entire design to be retargeted to the new process node even when only one core in the design requires it due to the performance requirements.

2.5D integration can be used for SM to overcome some of the issues discussed in the previous section with 2D SM. The idea is to use a secure interposer for the hidden BEOL wires and one or more FEOL chips produced at potentially untrusted foundries. Adapting 2D SM for security, overcoming the SM compatibility issues between the two foundries requires modifying the design process as discussed in the previous section. These issues are easily overcome in the *2.5D SM* method since

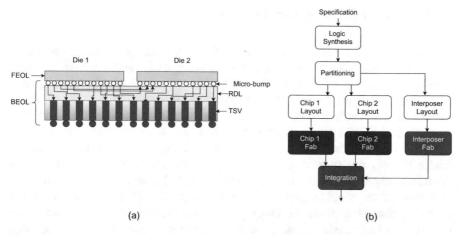

Fig. 1.9 2.5D integration. (**a**) Structure of a 2.5D IC, (**b**) SM process for 2.5D ICs

the chiplets and the interposer can all be manufactured at separate foundries. Since the IO buffers driving the micro-bumps usually operate at a much higher voltage compared to the internal logic, achieving timing closure for each chiplet is much easier. Further, the micro-bumps connecting the chiplets to the interposer are much larger than the vias used to connect the FEOL and BEOL parts in 2D ICs. This helps eliminate the PDK incompatibility issues and ease the process of interposer and chiplet bonding.

Figure 1.9b shows the SM design process when using 2.5D fabrication. This assumes a single chip with selected BEOL signals to be implemented in a secure interposer. Splitting the FEOL design into multiple chiplets and fabricating them at separate untrusted foundries add additional security since no one foundry has access to the complete FEOL netlist. Design partitioning can be accomplished manually at the architecture level or automatically at a lower level using automated netlist partitioning methods [55] if the target foundries are PDK compatible. In the event multiple chiplets need to be fabricated at different FEOL foundries each with its own PDK, complicated library remapping steps will be necessary after logic synthesis.

Locations of the micro-bumps should be determined after considering both the signal routing within the chiplets and the routing of the wires within the interposer. If only the chiplet designs are considered, performance may suffer from the long wires in the interposer. Place and route tools used for the chiplets usually do not consider how wires within the interposer are routed. This may not be a significant issue if these BEOL signals are not on the critical paths and have sufficient slacks. However, if these interposer signals are on the critical paths, there would be a significant performance penalty. Hence, the locations of the micro-bumps must be determined taking into account the connectivity inside the interposer before place and route is performed in each chiplet.

1.11 SM for 3D Integrated Circuits

3D integration refers to the manufacturing of ICs by stacking silicon dies or wafers and interconnecting them vertically. Various 3D integration methods exist and newer methods are emerging [56–58]. 3D stacked ICs are constructed by stacking dies and connecting them through TSV structures. Both face-to-face (F2F) and face-to-back (F2B) bonding methods can be used as shown in Figs. 1.10 and 1.11 respectively. In F2F bonding, metal wires can be directly bonded through and TSVs may not be necessary. Monolithic 3D ICs are built on a single wafer. Multiple device layers are connected using regular vias between metal layers. Monolithic 3D ICs are expected to continue to provide Moore "equivalent" integration benefits over the next decade. Compared to 2D integration, 3D integration offers smaller footprint, shorter interconnect, lower power consumption, and higher bandwidth between vertical device layers. In addition, similar to 2.5D ICs, 3D integration allows heterogeneous processes or process nodes to be used for the circuit layers. All of these translate into significant cost and performance advantages in favor of 3D integration.

Regardless of which 3D integration technology is used, SM for 3D circuits is similar to that for 2.5D circuits. A design can be partitioned into multiple chiplets. BEOL signals can be separated from one of more chiplets. FEOL parts of the chiplets can be produced at open foundries using the best processes suited for each of them. These parts can then be composed along with the BEOL interconnects using 3D integration. The difficulties involved with the interposer signal routing in 2.5D ICs are eliminated in 3D ICs. Surface area of the IC tends to be smaller in 3D integration compared to 2.5D integration. It is also possible to include an RDL in 3D ICs to effectively hide the BEOL signals. Dofe et al. [59], Xie et al. [60], Knechtel et al. [61], and others discussed the benefits, challenges, and new vulnerabilities of 3D ICs.

Fig. 1.10 F2F 3D integration

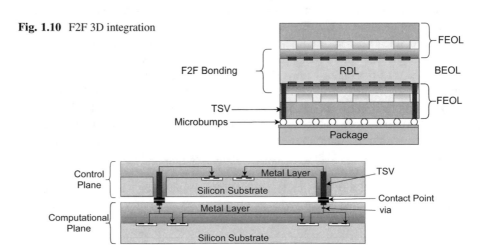

Fig. 1.11 A secure architecture using F2B 3D integration (based on [65])

Potential security benefits of 3D ICs have been noted by many researchers, for example [59–69], and include the following: (1) F2F 3D integration leaves the outside surface of the IC blank except for the IO pads. Reverse engineering of 3D ICs is challenging since even etching of the substrates is hard. Stacking conceals the circuitry and makes access difficult. (2) Critical wires and gates can be included in the *trusted tiers* fabricated at trusted foundries and the rest of the design can remain in *untrusted tiers* fabricated at potentially untrusted foundries. (3) In addition, the design can be partitioned into multiple chips and the interconnects between the chips can be randomized and hidden in the RDLs. Even when each chip and RDL are fabricated at an untrusted foundry, the overall design can still remain secure. (4) Run-time functional monitors, activity, power, temperature or aging sensors and other apparatus which are required for trojan detection and similar security related purposes can be implemented in the trusted tiers. (5) Various emerging technologies and devices [32] can be easily added and hidden among the stacked layers.

Exploiting these benefits of 3D integration, several authors proposed secure architectures. Huffmire et al. [63], Bilzor [70], and Valamehr et al. [64, 65] proposed the architecture, shown in Fig 1.11, using a processing tier fabricated at an untrusted foundry, a control tier fabricated at a secure foundry, and integrating them at a secure 3D integration facility. The secure controller can be used to monitor the processing circuitry and thwart a variety of attacks. Narasimhan et al. [71] proposed the use of current monitors, hidden in the trusted tiers, to detect trojans. Xie et al. [72] and Shi et al. [73] proposed an authentication method using specially designed circuitry to ensure the absence of trojans. The special circuitry, to be discussed in Sect. 5.6, can be hidden in the trusted tier. As an alternative to SM based protection, Dofe et al. [59] proposed a network-on-chip (NoC) based shielding plane between two untrusted tiers to thwart reverse engineering attacks.

1.12 Attacks Against SM

Ever since SM was considered for enhancing IC security, researchers began identifying potential vulnerabilities in SM that can be exploited by attackers for various malicious purposes. In this section, we summarize the proposed attack models some of which will be discussed in detail in this book. Unless stated otherwise, assume that we are dealing with a digital logic circuit implemented in static CMOS using the standard cell design methodology. It is this class of designs that are widely studied by researchers from a security perspective.

1.12.1 Attacker's Location

Vast majority of attacks against SM assume that the attacker is located at the FEOL foundry and that the BEOL foundry is trusted. While each attack targets 2D, 2.5D, or 3D ICs, many attacks can be adapted, with minor changes, to any split IC. From a security viewpoint, the SM method assumes that the BEOL information is not available to potential attackers. On the other hand, an attacker, located at the FEOL untrusted foundry should be assumed to have access to the entire FEOL information and is free to use it to infer any information regarding the missing BEOL signals. Most SM methods assume that at least the lower one or two levels of metal are a part of the FEOL structure and that the split takes place at some level above M1 or M2. This implies that the attacker can extract most, if not all, of the cells in the netlist since intra-cell routing among the devices is done using polysilicon, diffusion and one or two lower levels of metal wiring. In addition, several critical signal nets, especially short distance nets, could be fully or partly extracted. If power rails or global signals such as the clock signals are routed in lower levels of metallization, those nets can also be fully or partly extracted. In addition, the primary input and output signals which are connected to input and output pads are also assumed to be identifiable. Starting with this initial netlist and making use of any indirect hints from this netlist or the FEOL layout, the attacker's goal is to recover some or all of the missing BEOL nets thereby reconstructing a part or all of the original netlist or its functional equivalent.

1.12.2 Attacker's Objective

Attacks proposed to date have considered one of the following objectives:

1. *Reverse Engineering:* An attacker located at the FEOL foundry, having access to the FEOL layout and the FEOL netlist, may be interested in reverse engineering the entire netlist by reconstructing the missing BEOL signals. Due to the missing BEOL nets, the extracted FEOL netlist would contain gates with dangling input or output terminals. The attacker's task is to determine which output terminal drives which input terminal to correctly complete the logic circuit. The difficulty of the problem of recovering the missing BEOL signals can be estimated by the number of ways in which those dangling output terminals can be connected to the dangling input terminals. We assume that the circuit implemented is a combinational logic circuit for ease of analysis.

   Assume that the number of dangling input terminals is n and the number of dangling output terminals is m. Assume that every input terminal should have one and only one output terminal as a driver. Since each of the n input terminals can be driven by any of the m output terminals, the total number of possible connection patterns p is m^n. Since each connection pattern represents a potential logic design, the attacker needs to determine which of these p netlists represents the correct logic design. This assumes that each output terminal can fanout to drive up to n input terminals, that is, the fanout factor f for any BEOL net is $\leq n$. On the other hand, if each output terminal is assumed to drive exactly one input terminal, that is $f = 1$ for every missing net and $n = m$, then the number of possible netlists $p = n! = m!$.

Either extreme represents a daunting space of logic designs for the attacker. Of course, it is possible to make other assumptions on the fanout factor (for example, the fanout factor f for any net is bounded by a small constant) resulting in a design space between those two extremes.

2. *Hardware Trojan (HT) Insertion:* An attacker at the FEOL foundry may be interested in inserting a trojan at a location which has specific structural or functional characteristics. For example, the attacker may be looking for the output node of a cryptographic unit that satisfies a certain primitive function such as a DES s-box. As another example, the attacker may be looking for the output node of a chain of cells of a certain type. Given the FEOL layout and netlist, the attacker needs to identify the location(s) that match the target function or structure. This form of attack is also called a *layout recognition attack* since the attacker is attempting to recognize a location in the layout which matches a particular functional or structural feature.

   Imeson et al. [74] proposed the notion of k-*security* to capture the difficulty of layout recognition attacks. A location (output of a gate) in the layout is said to be k-*secure* if it cannot be distinguished from at least $k - 1$ other locations. In other words, the attacker faces at least k sites in the layout which match the function or structure he is interested in. For example, if all inter-cell nets are delegated to BEOL, then the k-security for any cell type is same as the number of cells of that type in the layout. The probability that the attacker then inserts the trojan at the correct location is $\leq 1/k$. The attacker either needs to pick one of these locations randomly and risk incorrect trojan insertion or insert trojans at multiple matching locations and risk easy detection by simple power measurements. k-security will be formally defined in Sect. 5.1.

1.12.3 Attack Methodology

Three broad classes of attack methodologies have been identified and discussed below:

1. *Design Constraint Based Attacks:* Attackers can exploit hints derivable from the layout such as cell and terminal types, relative locations of cells and terminals, directions of dangling wires and the layers where the dangling edges are located, drive strengths of the cells, etc. In addition, hints from commonly used EDA (Electronic Design Automation) tools and well-known design automation algorithms can be exploited. These hints are formulated as constraints on the structure of the design being reverse engineered such that a constraint-satisfying netlist is produced by the attack. Attacks based on this approach are called *design constraint based attacks*. It is possible to eliminate some of the connection

patterns using logic or layout hints. A wide variety of hints can be exploited. Some example types are as follows:

(a) *Logic Hints:* For example, in digital designs, combinational cycles are not allowed except in certain special circuits such as ring oscillators. These and other localized interconnect cycles such as those in static memory cells are implemented using FEOL layers. Hence, any net recovery which would introduce combinational cycles would be disallowed by the attacker. This would eliminate a number of possible netlists from p.

(b) *Layout Hints:* The attacker, located at the untrusted foundry, may have access to the FEOL netlist as well as the FEOL layout. In this case, additional hints can be obtained from the layout. For example, a gate with a low output resistance is likely to drive a large cell or drive multiple cells. A small cell is likely to drive a small cell in the near vicinity.

(c) *Cell Type Hints:* In high-performance designs, low threshold voltage (LVT), ultra low threshold voltage (ULVT), and extreme low threshold voltage (ELVT) cells are widely used for performance optimization in place of standard threshold voltage (SVT) cells. However, their use is usually limited to critical paths and specialized high-speed subcircuits. In addition, these cells are usually placed in close proximity and connected to each other in lower levels of metal. These considerations can be used to narrow the space of viable netlist candidates for recovery.

2. *Satisfiability Based Attacks:* Exploitable design hints represent educated guesses on the part of the attacker and usually provide an incomplete set of constraints for the attack. Simple and effective defense techniques can be used to mitigate the potential information leaks due to such guess work. Hence, design constraint based attacks, while relatively fast and can recover several missing BEOL signals, usually fall short of recovering all of the missing BEOL signals. Instead of depending on design constraints, *satisfiability* (SAT) based attacks use a black-box identification approach in which the attack is formulated as a Boolean expression such that any satisfying solution to the expression yields a successful recovery. These attacks benefit from the advances in Boolean satisfiability checking.

   Attacks can be also be categorized as two types based on oracle usage: *oracle-guided* and *oracle-less*. Oracle-guided attacks assume that the attacker is in a position to determine the output vectors of the IC given the inputs vectors, i.e. a stimulus-response oracle is available. This may be in the form of a working IC, possibly a functional equivalent fabricated at an older technology node, or a high level simulation model obtained in the open market if one is available or stolen from the design house. The attacker may be able to construct a functional model from the publicly available information such as the algorithms implemented in the IC. Oracle-guided attacks are quite potent but make the strong assumption that an oracle is available to the attacker. Oracle-less attacks assume that no such functional equivalent is available. Design constraint based attacks are usually oracle-less attacks whereas the SAT based attacks tend to be oracle-guided.

While they are quite effective in recovering the designs, satisfiability based attacks require excessive computer time and memory resources. This makes them viable only for small to medium scale designs at present.

3. *Hybrid Attacks:* Hybrid methods use satisfiability formulations but exploit design hints to significantly reduce the attack time and memory requirements. These combine the correctness of the SAT methods with the efficiency of the design constraint based methods.

In this book, design constraint based attacks will be discussed in Chap. 2 and defense methods to protect from these attacks will be discussed in Chap. 3. Satisfiability based attacks and hybrid attacks will be discussed in Chap. 4 and defense techniques to protect from such attacks will be discussed in Chap. 5.

1.12.4   Validation of the Recovered Design

In reverse engineering attacks, once a candidate design which includes the FEOL netlist and the reconstructed BEOL nets is recovered, the attacker needs to ensure that the candidate design is accurate with a high degree of confidence. This requires that the attacker simulate or emulate the candidate design using a sufficiently large number of input sequences and compare the results with those produced by an oracle suitable for this purpose. For combinational logic circuits with i inputs, exhaustive validation requires the application of 2^i vectors. For sequential circuits with i inputs and s internal state variables, exhaustive testing in general requires exercising the 2^{s+i} transitions. Clearly, exhaustive validation is intractable.

The attacker may instead choose to validate the recovered candidate design using a limited number of input sequences and decide whether to accept the recovered netlist or generate another alternative. In addition, if the error rate is sufficiently low, then the attacker might be satisfied with the recovery and assume that most of the recovered BEOL nets are correct. We will discuss various metrics used by researchers to determine the effectiveness of attacks and defenses in Chaps. 2–5.

1.12.5   Benchmarks

SM attacks and defense methods have been evaluated using several benchmark suites from a variety of sources. These benchmarks, ranging from small combinational logic circuits to relatively large industrial layouts, and their sources are summarized in Appendix A. Review of this appendix can help put the results discussed in the subsequent chapters in perspective.

1.13 Summary

Split manufacturing offers multiple security benefits in the context of the geographically dispersed IC design and fabrication processes which are subject to various uncertainties, vulnerabilities, and risks. While several early demonstrations support these benefits at minimal PPA costs, further large-scale evaluations of the cost-benefit tradeoff are needed before practical adoption of SM can be considered. Meanwhile, researchers have developed multiple attack methods against split designs and have also proposed defense techniques to thwart these attacks. These attacks and defense methods should be studied closely to determine their impact and practical significance. We have introduced the attackers' objectives and the general methods used. The rest of this book is concerned with the details of some of these attacks and defenses.

References

1. R.J. Baker, *CMOS: Circuit Design, Layout, and Simulation*, 4th edn. (Wiley, London, 2019)
2. N. Weste, D. Harris, *CMOS VLSI Design: A Circuits and Systems Perspective*, 4th edn. (Pearson Education India, 2010)
3. J. Rabaey, A. Chandrakasan, B. Nikolic, *Digital Integrated Circuits*, 2nd edn. (Pearson Education, 2002)
4. C. Alpert, D. Mehta, S. Sapatnekar, *Handbook of Algorithms for Physical Design Automation* (CRC Press, Boca Raton, 2008)
5. L. Lavagno, I.L. Markov, G.E. Martin, L.K. Scheffer, *Electronic Design Automation for IC System Design, Verification, and Testing* (CRC Press, Boca Raton, 2017)
6. L. Scheffer, L. Lavagno, G. Martin, *EDA for IC implementation, Circuit Design, and Process Technology* (CRC Press, Boca Raton, 2006)
7. S. Sait, H. Youssef, *VLSI Physical Design Automation: Theory and Practice* (World Scientific, Singapore, 1999)
8. C. Auth, A. Aliyarukunju, M. Asoro, D. Bergstrom, V. Bhagwat, J. Birdsall, N. Bisnik, M. Buehler, V. Chikarmane, G. Ding, Q. Fu, H. Gomez, W. Han, D. Hanken, M. Haran, M. Hattendorf, R. Heussner, H. Hiramatsu, B. Ho, S. Jaloviar, I. Jin, S. Joshi, S. Kirby, S. Kosaraju, H. Kothari, G. Leatherman, K. Lee, J. Leib, A. Madhavan, K. Maria, H. Meyer, T. Mule, C. Parker, S. Parthasarathy, C. Pelto, L. Pipes, I. Post, M. Prince, A. Rahman, S. Rajamani, A. Saha, J. Dacuna Santos, M. Sharma, V. Sharma, J. Shin, P. Sinha, P. Smith, M. Sprinkle, A. S. Amour, C. Staus, R. Suri, D. Towner, A. Tripathi, A. Tura, C. Ward, and A. Yeoh, "A 10 nm high performance and low-power CMOS technology featuring 3rd generation FinFET transistors, Self-Aligned Quad Patterning, contact over active gate and cobalt local interconnects, in *Technical Digest—International Electron Devices Meeting, IEDM* (Institute of Electrical and Electronics Engineers, Piscataway, 2018), pp. 1–29
9. CMOS-chip structure in 2000s (en)—Back end of line—Wikipedia. https://en.wikipedia.org/wiki/Back_end_of_line#/media/File:Cmos-chip_structure_in_2000s_(en).svg
10. W.J. Greig, *Integrated Circuit Packaging, Assembly and Interconnections* (Springer, Berlin, 2007)
11. R. Doering and Y. Nishi, *Handbook of Semiconductor Manufacturing Technology*, 2nd edn. (CRC Press, Boca Raton, 2007)
12. H. Xiao, *Introduction to Semiconductor Manufacturing Technology*, 2nd edn. (SPIE Press, 2012)

13. J. Lienig, J. Scheible, *Fundamentals of Layout Design for Electronic Circuits* (Springer, Berlin, 2020)
14. G.S. May, C.J. Spanos, *Fundamentals of Semiconductor Manufacturing and Process Control* (IEEE, Piscataway, 2006)
15. D. Brock, *Understanding Moore's Law: Four Decades of Innovation* (Chemical Heritage Foundation, Philadelphia, 2006)
16. International Roadmap for Devices and Systems (IRDS^{TM}) 2018 Edition—IEEE IRDS^{TM}. https://irds.ieee.org/editions/2018
17. J.P. Colinge, A. Chandrakasan, *FinFETs and Other Multi-Gate Transistors* (Springer, Berlin, 2008)
18. B.K. Kaushik, *Nanoelectronics: Devices, Circuits and Systems* (Elsevier, Amsterdam, 2018)
19. P. Mishra, S. Bhunia, M. Tehranipoor, *Hardware IP Security and Trust* (Springer, Berlin, 2017)
20. icinsights.com, Intel-To-Keep-Its-Number-One-Semiconductor-Supplier-Ranking-In-2020 (2020). https://www.icinsights.com/news/bulletins/Intel-To-Keep-Its-Number-One-Semiconductor-Supplier-Ranking-In-2020/
21. TrendForce, Press Center | TrendForce - Market research, price trend of DRAM, NAND Flash, LEDs, TFT-LCD and green energy, PV (2020). https://www.trendforce.com/presscenter/news/20200831-10463.html
22. Evertiq—Global top ten foundries for 2Q19 perform less-than-expected (2019). https://evertiq.com/news/46398
23. K. Xiao, D. Forte, Y. Jin, R. Karri, S. Bhunia, M. Tehranipoor, Hardware Trojans: lessons learned after one decade of research. ACM Trans. Des. Autom. Electron. Syst. **22**(1), 1–23 (2016)
24. S. Bhunia, M.S. Hsiao, M. Banga, S. Narasimhan, Hardware trojan attacks: threat analysis and countermeasures. Proc. IEEE **102**(8), 1229–1247 (2014)
25. S. Bhunia, M.M. Tehranipoor, *The Hardware Trojan War: Attacks, Myths, and Defenses* (Springer, Berlin, 2017)
26. S. Bhunia, S. Ray, S. Sur-Kolay, *Fundamentals of IP and SoC Security: Design, Verification, and Debug* (Springer, Berlin, 2017)
27. U. Guin, K. Huang, D. Dimase, J.M. Carulli, M. Tehranipoor, Y. Makris, Counterfeit integrated circuits: a rising threat in the global semiconductor supply chain. Proc. IEEE **102**(8), 1207–1228 (2014)
28. M. Tehranipoor, U. Guin, D. Forte, *Counterfeit Integrated Circuits: Detection and Avoidance* (Springer, Berlin, 2015)
29. E. Oriero, S.R. Hasan, Survey on recent counterfeit IC detection techniques and future research directions. Integration **66**, 135–152 (2019)
30. S.E. Quadir, J. Chen, D. Forte, N. Asadizanjani, S. Shahbazmohamadi, L. Wang, J. Chandy, M. Tehranipoor, A survey on chip to system reverse engineering. ACM J. Emer. Technol. Comput. Syst. **13**(1), 1–34 (2016)
31. R. Torrance, D. James, The state-of-the-art in semiconductor reverse engineering, in *Proceedings—Design Automation Conference* (2011), pp. 333–338
32. M. Tehranipoor, D. Forte, G.S. Rose, S. Bhunia, *Security Opportunities in Nano Devices and Emerging Technologies* (CRC Press, Boca Raton, 2017)
33. J. Knechtel, S. Patnaik, O. Sinanoglu, Protect your chip design intellectual property: an overview, in ACM International Conference Proceeding Series, vol. Part F1481 (2019), pp. 211–216
34. K. Shamsi, M. Li, K. Plaks, S. Fazzari, D.Z. Pan, Y. Jin, IP protection and supply chain security through logic obfuscation: A systematic overview. ACM Trans. Design Autom. Electron. Syst. **24**, 1–36 (2019)
35. J.P. Baukus, L.W. Chow, R.P. Cocchi, B.J. Wang, Method and apparatus for camouflaging a standard cell based integrated circuit with micro circuits and post processing, U.S. Patent, 20,120,139,582, 2012

36. N. Rangarajan, S. Patnaik, J. Knechtel, R. Karri, O. Sinanoglu, S. Rakheja, Opening the doors to dynamic camouflaging: harnessing the power of polymorphic devices. IEEE Trans. Emerg. Top. Comput. 1–19 (2020)

37. D. Forte, S. Bhunia, M.M. Tehranipoor, *Hardware Protection Through Obfuscation* (Springer, Berlin, 2017)

38. S.E. Hong, *Side Channel Attacks* (MDPI-Multidisciplinary Digital Publishing Institute, 2019)

39. L. Giancane, *Side-Channel Attacks and Countermeasures: Design of Secure IC's Devices for Cryptographic Applications* (LAP LAMBERT Academic Publishing, 2012)

40. J. Ambrose, A. Ignjatovic, S. Parameswaran, *Power Analysis Side Channel Attacks: The Processor Design-level Context* (VDM Verlag Dr. Müller, 2010)

41. M. Tehranipoor, H. Salmani, X. Zhang, *Integrated Circuit Authentication: Hardware Trojans and Counterfeit Detection* (Springer, Berlin, 2014)

42. M. Kaya, *Electronic Waste and Printed Circuit Board Recycling Technologies*, ser. The Minerals, Metals & Materials Series (Springer, Berlin, 2019)

43. S. Bhunia, M. Tehranipoor, *Hardware Security: A Hands-on Learning Approach* (Morgan Kaufmann, Los Altos, 2018)

44. D. Mukhopadhyay, R.S. Chakraborty, *Hardware security: Design, Threats, and Safeguards* (CRC Press, Boca Raton, 2014)

45. R.W. Jarvis, M.G. McIntyre, Split Manufacturing Method for Advanced Semiconductor Circuits, U.S. Patent 2004/0102019 A1, 2004

46. R. Jarvis, M.G. McIntyre, Split Manufacturing Method for Advanced Semiconductor Circuits, U.S. Patent 2007 7195931 B2, 2007

47. K. Vaidyanathan, B.P. Das, E. Sumbul, R. Liu, L. Pileggi, Building trusted ICs using split fabrication, in *Proceedings of the 2014 IEEE International Symposium on Hardware-Oriented Security and Trust, HOST 2014* (2014), pp. 1–6

48. Trusted Integrated Chips (TIC) IARPA (2014). https://www.iarpa.gov/index.php/research-programs/tic?highlight=WyJ0cnVzdGVkIl0=

49. K. Vaidyanathan, R. Liu, E. Sumbul, Q. Zhu, F. Franchetti, L. Pileggi, Efficient and secure intellectual property (IP) design with split fabrication, in *Proceedings of the 2014 IEEE International Symposium on Hardware-Oriented Security and Trust, HOST 2014* (IEEE Computer Society, Washington, 2014), pp. 13–18

50. K. Vaidyanathan, B.P. Das, L. Pileggi, Detecting reliability attacks during split fabrication using test-only BEOL stack, in *Proceedings—Design Automation Conference* (Institute of Electrical and Electronics Engineers, Piscataway, 2014), pp. 1–6

51. C.E. McCants, P. Manager, Intelligence Advanced Research Projects Activity (IARPA) IARPA Trusted Integrated Chips (TIC) Program, Tech. Rep., 2016. https://www.ndia.org/-/media/sites/ndia/meetings-and-events/divisions/systems-engineering/past-events/trusted-micro/2016-august/mccants-carl.ashx

52. G. Keskin, J. Proesel, L. Pileggi, Statistical modeling and post manufacturing configuration for scaled analog CMOS, in *Proceedings of the Custom Integrated Circuits Conference* (2010), pp. 1–4

53. B. Hill, R. Karmazin, C.T.O. Otero, J. Tse, R. Manohar, A split-foundry asynchronous FPGA, in *Proceedings of the Custom Integrated Circuits Conference*, no. Section V. (IEEE, Piscataway, 2013), pp. 1–4

54. X. Zhang, J.K. Lin, S. Wickramanayaka, S. Zhang, R. Weerasekera, R. Dutta, K.F. Chang, K.J. Chui, H. Y. Li, D. S. Wee Ho, L. Ding, G. Katti, S. Bhattacharya, D.L. Kwong, Heterogeneous 2.5D integration on through silicon interposer. Appl. Phys. Rev. 2(2), 021308–021308 (2015)

55. L.T. Wang, Y.W. Chang, K.T. Cheng, *Electronic Design Automation: Synthesis, Verification, and Test* (Morgan Kaufmann, Los Altos, 2009)

56. K. Sakuma, *3D Integration in VLSI Circuits: Implementation Technologies and Applications* (CRC Press, Boca Raton, 2018)

57. V.F. Pavlidis, I. Savidis, E.G. Friedman, *Three-Dimensional Integrated Circuit Design* (Newnes, 2017)

58. Y. Xie, J.J. Cong, S. Sapatnekar, *Three Dimensional Integrated Circuit Design*, ser. Integrated Circuits and Systems (Springer, Berlin, 2010)
59. J. Dofe, Q. Yu, H. Wang, E. Salman, Hardware security threats and potential countermeasures in emerging 3D ICs, in *Proceedings of the ACM Great Lakes Symposium on VLSI, GLSVLSI*, vol. 18 (2016), pp. 69–74
60. Y. Xie, C. Bao, C. Serafy, T. Lu, A. Srivastava, M. Tehranipoor, Security and vulnerability implications of 3D ICs. IEEE Trans. Multi-Scale Comput. Syst. **2**(2), 108–122 (2016)
61. J. Knechtel, O. Sinanoglu, I.A.M. Elfadel, J. Lienig, C.C. N. Sze, Large-scale 3D chips: challenges and solutions for design automation, testing, and trustworthy integration. IPSJ Trans. Syst. LSI Des. Methodol. **10**, 45–62 (2017)
62. Tezzaron, 3D-ICs and Integrated Circuit Security, Tech. Rep., 2008. http://www.tezzaron.com/about/PhotoAlbum/Products/3D_Sensor.html
63. T. Huffmire, T. Levin, M. Bilzor, C.E. Irvine, J. Valamehr, M. Tiwari, T. Sherwood, R. Kastner, Hardware trust implications of 3-D integration, in *Proceedings of the 5th Workshop on Embedded Systems Security, WESS '10* (2010), pp. 1–10
64. J. Valamehr, M. Tiwari, T. Sherwood, R. Kastner, T. Huffmire, C. Irvine, T. Levin, Hardware assistance for trustworthy systems through 3-D integration, in *Proceedings—Annual Computer Security Applications Conference, ACSAC* (2010), pp. 199–210
65. J. Valamehr, T. Sherwood, R. Kastner, D. Marangoni-Simonsen, T. Huffmire, C. Irvine, T. Levin, A 3-D split manufacturing approach to trustworthy system development. IEEE Trans. Comput. Aided Des. Integr. Circuits Syst. **32**(4), 611–615 (2013)
66. P. Gu, S. Li, D. Stow, R. Barnes, L. Liu, Y. Xie, E. Kursun, Leveraging 3D technologies for hardware security: Opportunities and challenges, in *Proceedings of the ACM Great Lakes Symposium on VLSI, GLSVLSI*, vol. 18-20-May-, 2016, pp. 347–352
67. J. Dofe, P. Gu, D. Stow, Q. Yu, E. Kursun, Y. Xie, Security threats and countermeasures in three-dimensional integrated circuits, in *Proceedings of the ACM Great Lakes Symposium on VLSI, GLSVLSI*, vol. Part F1277, 2017, pp. 321–326
68. Y. Xie, C. Bao, A. Srivastava, 3D/2.5D IC-based obfuscation, in *Hardware Protection Through Obfuscation* (Springer, Berlin, 2017), pp. 291–314
69. J. Knechtel, S. Patnaik, O. Sinanoglu, 3D integration: another dimension toward hardware security, in *2019 IEEE 25th International Symposium on On-Line Testing and Robust System Design, IOLTS 2019* (Institute of Electrical and Electronics Engineers, Piscataway, 2019), pp. 147–150
70. C. Michael Bilzor, 3D execution monitor (3D-EM): using 3D circuits to detect hardware malicious inclusions in general purpose processors, in *Proceedings of the Sixth International Conference on Information Warfare Security* (2011), pp. 288–300
71. S. Narasimhan, W. Yueh, X. Wang, S. Mukhopadhyay, S. Bhunia, Improving IC security against trojan attacks through integration of security monitors. IEEE Design Test Comput. **29**(5), 37–46 (2012)
72. K. Xiao, D. Forte, M.M. Tehranipoor, Efficient and secure split manufacturing via obfuscated built-in self-authentication, in *Proceedings of the IEEE International Symposium on Hardware-Oriented Security and Trust* (2015), pp. 14–19
73. Q. Shi, K. Xiao, D. Forte, M.M. Tehranipoor, Obfuscated built-in self-authentication, in *Hardware Protection through Obfuscation* (Springer, Berlin, 2017), pp. 263–289
74. F. Imeson, A. Emtenan, S. Garg, M.V. Tripunitara, Securing computer hardware using 3D integrated circuit (IC) technology and split manufacturing for obfuscation, in *Proceedings of the 22nd USENIX Security Symposium* (USENIX Association, Berkeley, 2013), pp. 495–510

Chapter 2
Design Constraint Based Attacks

Abstract Attacks against SM which exploit information inferred from the FEOL layout combined with the knowledge of common EDA flows are termed *design constraint based attacks*. Examples of such information include cell types, terminal positions, distance between terminals, directions of wires, and loads offered by various cells. In this chapter, we discuss several design constraint based attacks for reverse engineering and trojan insertion. Specifically, for reverse engineering, we discuss attacks based on proximity and notions of extended proximity, the network flow model, and *machine learning*. For trojan insertion, we discuss attacks based on simulated annealing, proximity based mapping followed by net based pruning, and structural pattern matching.

2.1 General Approach

Design constraint based attacks exploit *hints* obtained from the FEOL layout and netlist combined with the knowledge of design practices and EDA flows. Jagasivamani et al. [1] postulated a generic design constraint attack process. In this iterative attack process, one or more of the following steps can be performed in each iteration to progressively reconstruct the missing BEOL signals:

1. *Cell Identification:* This is the process of identifying the basic functional cells used in the design. Since the basic cells are implemented in the device layers and lower levels of metal for intra-cell interconnect, cell layouts are available in the FEOL layers. Basic cells can include cells usually available in standard cell libraries to support synthesis flows and analog building blocks or other specialized IO circuits which have easily recognizable structures.
2. *Composition Analysis:* This step attempts to analyze the types of cells used in the design to infer the missing connections. For example, mix of cells used (e.g. predominantly XOR gates to denote crypto circuits) and presence of anchor cells (e.g. adders to implement ALUs) yield clues about design.
3. *Proximity Analysis:* This step analyzes the layout for clues from well-known optimizations used in physical design by layout synthesis tools. Connected neigh-

© The Author(s), under exclusive license to Springer Nature Switzerland AG 2021
R. Vemuri, S. Chen, *Split Manufacturing of Integrated Circuits for Hardware Security and Trust*, https://doi.org/10.1007/978-3-030-73445-9_2

bors being placed close to each other to minimize wire length and potentially maximize performance, drive to load matching, direction of dangling wires, etc. are among the possible clues that help an attacker in recovering the correct BEOL connectivity.

4. *Regional Examination:* This step attempts to infer functionality based on connections to known or other inferred blocks. For example, common practices such as IO circuitry should be on the boundary of the chip, signal processing usually follows analog–digital conversion, etc. can be used to make module level inferences.

In this chapter, we will discuss several design constraint based attacks devised by researchers using these and similar principles.

2.2 Attack Evaluation Metrics

Several metrics can be defined to evaluate the effectiveness of the attack and defense methods. Some of the commonly used *metrics* are defined below:

1. *Attack Correctness (AC) or Correct Connection Rate (CCR):* An attack is successful if the attacker is able to correctly recover the hidden BEOL signals. *Attack Correctness* (AC) is defined as the percentage of the BEOL signals correctly recovered. This is computed as $|H' \cap H|/|H|$ where H is the set of hidden BEOL signals and H' is the set of recovered BEOL signals. AC is also called the *Correct Connection Rate* (CCR) and is usually reported as a percentage. If AC is 100%, then the circuit is completely reverse engineered. Note that the attacker cannot compute AC since H is unknown to him. However, researchers can determine this in their experimental studies since the original netlists are available to them.

2. *Incorrect Connection Rate (ICR):* This is defined as the percentage of incorrect connections between the BEOL and FEOL parts in the recovered circuit after an attack. ICR is the complement of CCR.

3. *Percentage of Netlist Recovered (PNR):* This is defined as the ratio of correctly inferred nets in the recovered design over the total number of nets in the original design. PNR is a direct measure of the correctness of the entire design rather than just the protected nets.

4. *Hamming Distance (HD):* The attacker may simulate the recovered netlist and compare the results with those produced by the actual IC or an oracle, if one is available. *Hamming Distance* (HD) is widely used to quantify the attack correctness in such a scenario. Hamming Distance, $HD(x, y)$, between two Boolean vectors x and y of the same size is defined as the number of positions in which they differ. Let $Y = F(X)$ be the Boolean function represented by the original logic circuit G. X and Y are vectors of Boolean input and

output variables, respectively. Let F' be the Boolean function represented by the recovered circuit G'. If x is a specific Boolean input vector then $HD(y, y')$, where $y = F(x)$ and $y' = F'(x)$, is the number of bit positions in which y and y' differ.

For a given input vector x, the *Normalized Hamming Distance*, NHD(x) is defined as the ratio $HD(F(x), F'(x))/|Y|$ where the denominator denotes the number of output variables. Given a series of input vectors $x_1, x_2, \ldots x_n$, *Average NHD* (ANHD) is defined as follows:

$$\text{ANHD} = \frac{\sum_{i=1}^{n} \text{NHD}(x_i)}{n} \tag{2.1}$$

For a sufficiently long sequence of input vectors, if ANHD is close to 0% (where F(X) is determined by the oracle used by the attacker and F'(X) is computed by simulating the recovered circuit) then the attack is assumed to be successful. On the other hand, ANHD of 100% also indicates a successful attack since this indicates that the vectors in the two sequences are bitwise inversions of each other [2, 3]. Two uncorrelated circuits will produce output sequences whose ANHD is in the vicinity of 50% [4, 5]. While the attacker tries to recover a circuit that yields ANHD in the vicinity of 0% or 100%, the goal of split manufacturing is to ensure that the ANHD stays as close as possible to 50% over arbitrarily long input vector sequences. This ensures that not only the attack is unable to recover the hidden nets but also that the recovered circuit and the original circuit are functionally uncorrelated.

While an attacker without access to a functional IC or model cannot evaluate ANHD, researchers have used the HD metric to evaluate the SM attack and defense methods. ANHD is simply referred to as the *HD metric*.

5. *Output Error Rate (OER):* The HD metric measures the *average* number of bit positions in which the output vectors of the original and recovered circuits differ. The *output error rate* (OER) is a direct measure of incorrect output instances and is defined as the number of incorrect outputs produced while applying a specified number of input vectors. Given the number of input vectors, the OER is usually reported as a percentage.

Several other metrics used in SM research will be introduced as needed.

2.3 Proximity Attack

By making certain reasonable assumptions about the design process, an attacker can discard many incorrect nets during the BEOL net recovery process. The assumptions are based on how commonly used design automation algorithms and design flows work [6, 7]. Rajendran et al. [2] proposed an attack named the *proximity attack* for combinational circuits. The attack exploited well-known heuristics used in physical design tools to correctly recover the BEOL nets in split manufactured combinational logic circuits.

2.3.1   Attack Assumptions

The proximity attack assumes that, during the design process, the BEOL nets are identified using a circuit bipartitioning algorithm. It is further assumed that the two partitions are placed and routed such that all of the cells and nets within each partition are delegated to the FEOL layers and the nets connecting the two partitions are delegated to the BEOL layers. The attacker, having access to the FEOL layout, is able to reverse engineer the logic circuits of the two partitions. Then, he is ready to mount the proximity attack. The attack is based on the following observations:

1. *Pin Direction:* An input pin (terminal) of a partition should be connected to an output pin of the other partition or an input port of the IC. Conversely, an output pin of a partition should be connected to an input pin of the other partition or an output port of the IC.
2. *Output-Input Pairs:* An output pin of a partition should only be connected to one input pin of another partition. This is based on the assumption that if the output pin had fanout into the other partition, then the fanout net would have been moved into the target partition in order to minimize the cutset (number of wires) between the two partitions. This assumption reduces each BEOL connection to a two terminal wire and avoids multi-terminal nets.
3. *Combinational Loops:* Since cycles are not allowed in combinational logic circuits, any connection between partitions that would introduce a cycle in the logic circuit is disallowed.
4. *Compact Placement:* Placement tools would have placed the cells in the two partitions and oriented them so as to minimize the total wire length. Long wires increase wire delays, reduce circuit performance, and increase power consumption. Accordingly, wire length minimization is a commonly used optimization goal during physical design. Hence, the two partition pins to be connected (by a hidden BEOL net) are likely to be in close proximity of each other. The proximity attack exploits this observation while prioritizing the nets to be recovered and hence the name.

2.3.2   Attack Algorithm

Algorithm 1 shows the proximity attack algorithm. In each iteration of the while loop, an unassigned input pin of a partition or an output port of the circuit is selected as the target pin (line 3). Then, a *list of candidate* pins is constructed. All of the unassigned output pins of the other partition or input ports of the circuit constitute

---

Algorithm 1: Proximity attack [2]

Input: FEOL Layers
Output: Netlist with BEOL connections
1 Reverse engineer FEOL layers and obtain the partitions;
2 **while** *unassigned partition pins or ports exist* **do**
3   Select any unassigned input pin or an output port as a target pin, t;
4      $C = $ CandidatePins(t);
5   Select candidate pin, $p \in C$ such that p is closest to t;
6   Connect t and p;
7   Update netlist;
8 **end**
9 **return** netlist;
10 —————————————————————————————————————
11 **CandidatePins** *(t)*

    Input: Target Pin t
    Output: Candidate Pins List C for t
12 $C = $ unassigned output pins of the other partition + unassigned input ports;
13 **for** *each pin $p \in C$* **do**
14   **if** *connecting t and p forms a combination cycle in the logic circuit* **then**
15          $C = C - \{p\}$;
16   **end**
17 **end**
18 **return** C;

---

the candidate pins. From this list, any pin which, if connected to the target pin, would introduce a cycle in the circuit is excluded (lines 13–16). From the resultant candidate pins, a pin which is closest to the target pin is selected (line 5). To determine closeness, rectilinear distance between the two pins, as measured from the FEOL layout, is used. The selected candidate pin and the target pin are connected. This connection represents a recovered BEOL net. This process continues until all partition inputs and circuit outputs are assigned.

Consider the example netlist in Fig. 2.1 which was partitioned into A and B and placed. Table 2.1 shows the coordinates of the unconnected pins and ports. Suppose BI3 is selected as the target pin. Candidate pins are {I1, I2, I5, AO1}. Among these, AO1 is the nearest pin; hence, BI3 is connected to AO1. Suppose AI2 is selected as the next target pin. Candidate pins are {I1, I2, I5, BO1, BO2}. Both BO1 and BO2 will be discarded since both form combinational cycles. Among the remaining pins, I2 is the closest and will be connected to AI2. The algorithm continues until all the remaining pins are connected.

2.3.3 Discussion

Following two-way partitioning, placement and routing on ISCAS-85 benchmarks, the proximity attack was able to recover 96% of BEOL connections correctly [2].

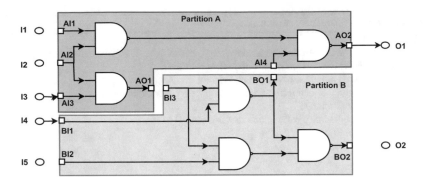

Fig. 2.1 Example to illustrate the proximity attack (based on [2])

Table 2.1 Example
coordinates of pins and IO
ports

IO port or pin	Coordinates
IO ports	
I1	(0,15)
I2	(0,10)
I5	(0,0)
O2	(21,3)
Partition A	
AI1	(2,15)
AI2	(2,10)
AI4	(14,9)
AO1	(7,6)
Partition B	
BI2	(2,0)
BI3	(9,6)
BO1	(14,7)
BO2	(19,3)

The proximity attack requires the two partitions to be clearly identified in the recovered FEOL circuit. This implicitly assumes that the splitting is done at a sufficiently high level metal layer such as M4. The attack is extremely fast and can potentially handle large circuits. However, the attack cannot guarantee correctness in terms of the AC metric. The attack was not evaluated using the HD metric.

The attack does not use any ordering heuristics to select the order of the target pins. Further, there is no tie-breaking heuristics to prioritize candidate pins that are equidistant from a target pins. The attack does not backtrack. Thus, once a target pin is connected to an incorrect candidate pin, there is no way to undo this decision, thereby subsequently forcing additional errors.

The proximity attack may fail to produce a complete netlist. This can happen when, for example, for a certain target pin, all of remaining candidate pins form combinational cycles due to an earlier incorrect connection. In this case, the target pin cannot be connected to any of its candidate pins. All of these features can contribute to lowering the attack correctness.

The attack is based on simplistic assumptions about the physical design algorithms. For example, placement algorithms attempt to minimize the total wire length. This implies that each pair of connected cells may not be in their optimal positions. Hence, the proximity heuristic could make incorrect decisions as it assumes that the placement locally minimizes the length of each inter-partition wire.

The proximity attack will have low success rate if the design processed used more sophisticated net lifting methods such as multiway partitioning or performance-driven partition, place and route methods. In such design flows, strict proximity based recovery has low probability of success.

2.4 Extended Proximities

The proximity attack prioritizes BEOL connectivity based on the "nearest neighbor" heuristic which is fast but *greedy*. The search neighborhood can be enlarged based on other considerations to increase the chances of successful BEOL recovery.

2.4.1 Types of Proximity

Attackers can define the search neighborhoods based on various notions of proximity. Magaña et al. [8, 9] proposed four such notions:

1. *Placement proximity:* This is defined as a square neighborhood centered around the location of the standard cell input or output pin connected to the dangling wire fragment. The *bounding box* (BB) area is set to be equal to (or slightly larger than) the average area of the BBs containing similar nets in typical designs. Alternatively, the average area of the BBs of the terminals of all uncut nets in the FEOL design can be measured and used by the attacker. For example, Fig. 2.2a shows a placement layer and several routing layers in which nets are routed. Figure 2.2b shows a BB denoting placement proximity around the output pin of a gate and its candidate pins in the BB. Note that, for convenience of drawing, the BB in the figure is not precisely centered on the considered pin.
2. *Routing proximity:* The is defined as a square area centered around a virtual pin (vpin) in the cut layer. Figure 2.2b shows an example routing proximity BB around a considered vpin and its candidate vpins identified in the BB. The attacker can determine this area to be the average area of the bounding boxes of connected vpin locations in the same layer based on measurements using typical designs. Note that this average area could be different in different layers. In general, this BB area increases for higher layers since wires become wider, making vpins of the same net to be spread apart. Although larger in area, these

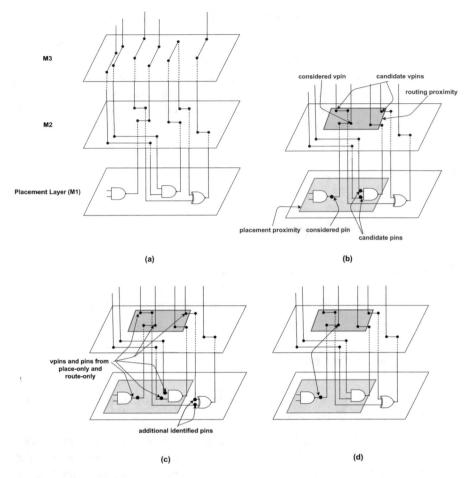

Fig. 2.2 Proximity definitions (based on [8]). (**a**) Placement layer and routing layers, (**b**) Placement proximity and routing proximity, (**c**) Union of placement and routing proximities, (**d**) Overlap of placement and routing proximities

BBs are likely to contain fewer vpins at higher levels due to the increased spacing between wires and increased wire widths. To take routing congestion into account, the BB around a vpin can be expanded uniformly in all directions until the ratio of the number of vpins in the BB to the BB area (vpin density) reaches the same value (or a desired target value) for all the vpins in the split level. This variation is named *crouting proximity*.

3. *Union of placement and routing proximities:* In this method, both placement and routing proximity BBs are determined. In addition to all the gate IO pins in the placement proximity, all of the placement pins (i.e. source or sink pins of gates) corresponding to all the vpins in the routing proximity are included in the search space. Figure 2.2c shows an example.

4. *Overlap of placement and routing proximities:* As in the previous case, both placement and routing proximity BBs are determined. However, the final search includes only the subset of the pins found in the placement proximity such that their corresponding vpins are included in the routing proximity. Figure 2.2d shows an example.

Magaña et al. [9] also discussed ways to expand these proximities for nets that have multiple vpins in the split level. Essentially, the BBs are expanded based on measuring average BB areas for multi-vpin nets (3 to 8 vpins per net) in different layers. An attacker can make such measurements on typical designs and use that data to adjust the search area.

2.4.2   Discussion

Magaña et al. [8] reported detailed experiments using five ISPD-2011 benchmarks to analyze the five variations of the proximity based search area identification: placement, routing, crouting, union of place and route, and overlap of place and route. A simple proximity attack which for each considered vpin selects the candidate vpin closet to it from the list of candidates in the search area is used for determining attack resilience. The methods are compared based on the average size, $E[SA]$, of the search area (in terms of the number of global cells) at each split level; the average size, $E[LS]$, of the candidate list; and the average ratio, $E[LS/SA]$, of LS to SA at each split level. $E[LS/SA]$ is a figure of merit (FOM); the higher the value, the more challenging the attack.

In terms of %match (the percentage of vpins for which the correct match is included in the candidate list) Magaña et al. found that crouting, routing, union, overlap and placement methods for 2-vpin nets at split level M2 resulted in 82.08%, 71.08%, 71.08%, 13.05%, and 12.94% matches respectively. Crouting has the highest $E[SA]$ which implies a higher chance of a larger candidate list and a higher possibility of the correct match in that list. This indicates the importance of routing congestion in defining proximity. The %match drops at higher levels as $E[LS]$ gets smaller due increased wire sizes.

When the proximity attack was used, the actual number of matches was low even for the crouting method which has a high %match. This indicates that the matching vpin is not necessarily the closest one to the considered vpin. Additional techniques to prune the candidate lists are essential to find the correct match. In terms of FOM, crouting and routing had higher values compared to the other methods and offer better security.

These experiments indicate that greedy proximity based attacks are unlikely to succeed in recovering the BEOL nets. Further, the enlarged neighborhoods covered by the extended proximity searches are more likely to include the correct matching vpins.

2.5   Network Flow Attack

The proximity attack, discussed in Sect. 2.3, is a fast but greedy heuristic. The procedure is *sequential*, making one connection at a time without backtracking. To overcome the limitations of the basic proximity attack, Wang et al. proposed an enhanced form, named the *network flow attack* [10, 11] based on modeling the BEOL net recovery problem as a network flow problem. The network flow attack does not require that the design be partitioned and works directly on the flattened netlists. Recall that in the proximity attack, once incorrect connections are introduced, the attack correctness reduces rapidly since more incorrect connections would be established in subsequent iterations. The network flow attack avoids making incorrect connections as much as possible by considering additional hints and simultaneously inferring multiple connections thereby dramatically increasing the attack correctness.

2.5.1   Attack Assumptions

The network flow attack assumes that the attacker, located at the untrusted FEOL foundry, has access to the technology library as well as the complete FEOL layout. This implies that he can obtain the structure and delay of the logic cells, wire capacitance, and *load capacitance*. In addition to the pin direction hint, absence of combinational loops, and proximity hint due to the compact placement assumption used in the proximity attack, the network flow attack uses the following additional hints:

1. *Load Capacitance Matching:* The attacker, having access to the physical characteristic data of the standard cells used in the design, knows the drive strength of each cell in terms of the maximum load capacitance it can drive. Similarly, the load offered by a cell at each of its input pins is also known. The attacker can exclude any input pin p which offers a load that exceeds the drive strength of another cell's output pin q from the candidate list of input pins that can be connected to q.
2. *Direction of Dangling Wires:* A wire may consist of several wire segments each of which is routed in a layer different from the others. These wire segments are connected by vias. Depending on the metal level selected for splitting, some of these wire segments may be in the FEOL layout and others in the BEOL layout. Thus, several wires in the FEOL layout would be "dangling." If a dangling wire from an output node p is pointing in a certain direction, then the attacker can exclude input nodes of cells in other directions from the candidate list of potential sink nodes for p. This is based on the assumption that automated routers route each wire from the source nodes aggressively moving toward the sink nodes in order to minimize the wire length. Figure 2.3 shows dangling wires from two

Fig. 2.3 Direction of
dangling wires (based on
[10])

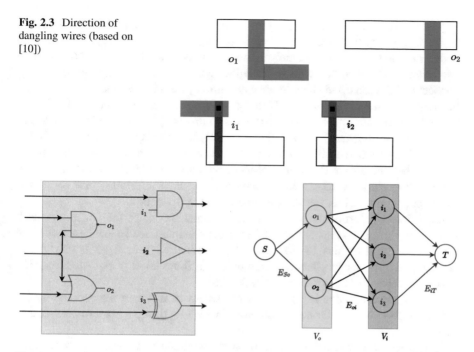

Fig. 2.4 Network flow model example (based on [10])

input and two output nodes. Pin i_2 has a dangling wire in the direction of o_2. Pin
o_1 has a dangling wire in the direction of i_2. Pin i_1 has a dangling wire in the
direction away from both o_1 and o_2.

3. *Timing Constraints:* The attacker, based on the knowledge or an educated guess
   of the clock frequency, can estimate the timing constraints on the signal paths.
   He can exclude any connection that would violate a timing constraint.

2.5.2   Attack Algorithm

The BEOL net recovery problem is modeled as a network flow problem. The reverse
engineered FEOL circuit F is represented as a directed graph $G = (V, E)$. V
consists of a unique source node S, a unique sink node T, a subset V_o which contains
a unique node for each output pin in F, and a subset V_i which contains a unique node
for each input pin in F. E consists of a subset E_{So} which contains an edge from S
to each node in V_o, a subset E_{iT} which contains an edge from each node in V_i to T,
and a subset E_{oi} which contains edges from each node in V_o to selected nodes in V_i
constructed based on the direction hint and the timing non-violation rule. Figure 2.4
shows an example. o_1 and o_2 are the unconnected output pins and i_1, i_2, and i_3 are
the unconnected input pins.

The network flow attack algorithm is shown as Algorithm 2. The algorithm begins by creating the nodes and source/sink edges in G (lines 1–11). Then for each node in V_o, an edge to every qualifying node in V_i is created (lines 12–16) based on two criteria: (1) pin o should have a dangling wire in the direction of pin i, and (2) connection between o and i should not cause timing violation. Slack at o is estimated by subtracting the *arrival time* (AT) at o from the required arrival time (RAT) at i. These times are determined based on the minimum operating frequency information available to the attacker. The slack does not include the delay between o to i. If this estimate is positive, then an edge is created between the corresponding nodes in the graph. When the arrival times are not available due to missing connections, AT at the primary inputs is used as a lower bound and the RAT at primary outputs is used as an upper bound. Thus, the slack estimate is an upper bound.

Capacitance information is used to define capacities c_e of the edges. For each source edge in E_{So}, its capacity is defined as the maximum load capacitance allowed for that output pin (lines 18–20). For each terminal edge in E_{iT}, its capacity is defined as the input capacitance of that input pin (lines 21–23). For the remaining edges, the capacities are set to ∞ (lines 24–26).

Cost w_e of each edge is defined as the length of the wire for connecting the corresponding pins. For the source and terminal edges, the cost is set to zero. (Lines 37–34)

The graph so constructed is then used to formulate and solve a network flow problem [12] as follows (line 36): For each edge $e \in E$, let x_e denote the flow through e. Constraints on the flow are as follows:

1. For each edge, flow should not exceed capacity.

$$\forall e \in E, \ x_e \leq c_e \tag{2.2}$$

2. For each node (except S and T), inflow must be equal to outflow.

$$\forall j \in V_o \cup V_i, \ \sum_{\forall i, (i,j) \in E} x_{(i,j)} = \sum_{\forall k, (j,k) \in E} x_{(j,k)} \tag{2.3}$$

3. Total inflow into T should equal the total sink capacity.

$$\sum_{\forall e \in E_{iT}} x_e = \sum_{\forall e \in E_{iT}} c_e \tag{2.4}$$

4. Total outflow from source S should equal the total inflow into the sink T.

$$\sum_{\forall e \in E_{So}} x_e = \sum_{\forall e \in E_{iT}} c_e \tag{2.5}$$

The optimization goal is to minimize the total flow cost which is the total weighted wire length:

Algorithm 2: Network flow attack [10]

Input: FEOL Layers L
Output: Netlist with BEOL connections F

1 Reverse engineer FEOL layers and obtain a netlist F;
2 Create graph $G = (V, E)$ where $V = \{S, T\}$ and $E = \emptyset$;
3 $V_o = \emptyset$; $V_i = \emptyset$; $E_{So} = \emptyset$; $E_{iT} = \emptyset$; $E_{oi} = \emptyset$
4 **for** *each unconnected output pin o of a cell in F* **do**
5 $\quad$ create a node v_o in V_o;
6 $\quad$ create an edge (S, v_o) in E_{So};
7 **end**
8 **for** *each unconnected input pin i of a cell in F* **do**
9 $\quad$ create a node v_i in V_i;
10 $\quad$ create an edge (v_i, T) in E_{iT};
11 **end**
12 **for** *each $v_o \in V_o$ and each $v_i \in V_i$* **do**
13 $\quad$ **if** *(output pin o is in the direction of input pin i in F)* **and** *(a connection between o*
$\quad\quad$ *and i would not cause a timing violation)* **then**
14 $\quad\quad$ add edge (v_o, v_i) to E_{oi};
15 $\quad$ **end**
16 **end**
17 $V = (V \cup V_o \cup V_i)$; $E = (E \cup E_{So} \cup E_{iT} \cup E_{oi})$;
18 **for** *each edge $e = (S, v_o) \in E_{So}$* **do**
19 $\quad$ c_e = load capacity of pin o;
20 **end**
21 **for** *each edge $e = (v_i, T) \in E_{iT}$* **do**
22 $\quad$ c_e = input capacitance of pin i;
23 **end**
24 **for** *each edge $e = (v_o, v_i) \in E_{oi}$* **do**
25 $\quad$ $c_e = \infty$;
26 **end**
27 **for** *each edge $e \in E$* **do**
28 $\quad$ **if** $e \in E_{So}$ **or** $e \in E_{iT}$ **then**
29 $\quad\quad$ $w_e = 0$
30 $\quad$ **end**
31 $\quad$ **if** $e \in E_{oi}$ **then**
32 $\quad\quad$ w_e = length of the wire connecting o and i;
33 $\quad$ **end**
34 **end**
35 **loop**
36 $\quad$ Formulate and solve the min-cost network flow problem using G;
37 $\quad$ Infer BEOL connections based on the solution and add them to F;
38 $\quad$ **if** *there is a cycle in F* **then**
39 $\quad\quad$ Delete the edge in G corresponding to the longest inferred BEOL connection;
40 $\quad$ **else**
41 $\quad\quad$ **return** F;
42 $\quad$ **end**
43 **forever**

$$\text{Minimize} \sum_{\forall e \in E} w_e . x_e \qquad\qquad (2.6)$$

The solution to this network flow problem yields flows through the edges E_{oi} from which the BEOL nets are inferred (line 37). Note that, in contrast to the proximity attack, the BEOL nets inferred can be multi-terminal nets.

To ensure that a cycle-free combinational circuit is extracted, an iterative approach is followed (lines 35–43). If there is a cycle in the extracted circuit, then the edge in E corresponding to the longest inferred correction is removed from the graph and the network flow problem is solved again. This process is repeated until a cycle-free netlist is extracted and returned (line 41).

Several algorithms exist to solve the network flow problem, for example, the Edmonds–Karp algorithm [12] which has a polynomial time complexity. At the end of each iteration, the connected BEOL pins can be removed to reduce the execution time. Network flow attack software is made available on the internet by Lang et al. [13].

2.5.3   Discussion

Wang et al. [10] demonstrated the effectiveness of the network flow attack on the ISCAS-85 and ITC-99 benchmarks. In their experiments, the network flow attack achieved 67% CCR while the proximity attack (Sect. 2.3) achieved less than 25% correct connections on average. OER for the circuits recovered by the proximity attack was 90% while the network flow attack reduced it to less than 50%. Time complexity of the attack algorithm is $O(V^2 E^2)$ where V and E are the number of vertices and edges in the network graph constructed.

While the attack is significantly more effective than the basic proximity attack, further improvements are possible. The process of eliminating cycles may converge toward an incorrect circuit. Various notions of proximity neighborhoods, discussed in Sect. 2.4, can be considered to introduce edges only between the nodes in the neighborhood in order to prune the size of the graph.

2.6   Machine Learning Attack

Both the proximity and the network flow attacks have incorporated the knowledge of the hints obtained from the design and design process into the attack algorithms. This knowledge comes from the attacker's experience and awareness. Alternatively, it is possible to capture such knowledge through explicit sampling of the appropriate features of several split designs and train a suitable model to capture the knowledge and use the model to attack a new FEOL design.

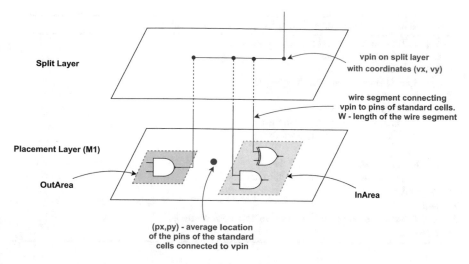

Fig. 2.5   Features related to vpins (based on [14])

Zhang et al. [14] and Zeng et al. [15] postulated such an attack using machine learning (ML) techniques. In this attack, a *classifier* is first trained using sample data generated from benchmarks and is validated. Then, the classifier is used to mount an attack on test cases to recover the missing BEOL connectivity.

For each benchmark, a layout is generated and cut at a selected split layer. The FEOL part is used as a challenge instance. The *split layer* is the via layer where the design is cut. Spilt layer k implies that metal layers up to and including k are part of the FEOL design and the layers above k are unavailable to the attacker at the FEOL foundry. The point where a net is broken in the split layer is referred to as the *virtual pin* or *vpin*. Each vpin designates the location of a via. Each vpin connects to one or more standard cell IO pins via a route fragment as shown in Fig. 2.5.

2.6.1   Classifier Development

The machine learning process is shown as Algorithm 3 and explained below:

1. *Feature Extraction:* For each vpin v, with coordinates (vx, vy), in the training design, the following layout *features* are computed:

   (a) Wire length W for the route fragment connecting v to the IO pins on the standard cells on the placement layer.
   (b) Location (px, py) where v connects to the placement layer. If it connects to multiple locations, the average location is determined.
   (c) Sum of the areas of the standard cells connecting to v through an input pin. This parameter is denoted by A_{in}.

Algorithm 3: Classifier training for proximity attack [15]

Input: Training Design Set $\mathcal{D}$, Split Layer k
Output: Trained Classifier C
1 **for** *each design $D \in \mathcal{D}$* **do**
2 $\quad$ Split D at layer k and locate vpin set V;
3 $\quad$ **for** *each $v \in V$* **do**
4 $\quad\quad$ Extract features of v;
5 $\quad\quad$ Generate a positive match for v;
6 $\quad\quad$ Generate a negative match for v;
7 $\quad$ **end**
8 $\quad$ **for** *each sample pair (v_1, v_2)* **do**
9 $\quad\quad$ Generate sample data for training;
10 $\quad$ **end**
11 **end**
12 Train classifier C using the combined sample data from $\mathcal{D}$;
13 **return** C;

(d) Sum of the areas of the standard cells connecting to v through an output pin. This parameter is denoted by A_{out}.

(e) Placement congestion (PC) defined as the pin density around the pin that connects to v.

(f) Routing congestion (RC) defined as the vpin density around v.

2. *Sample Generation:* For each vpin v, the correctly matching vpin is found. This pair constitutes a *positive* sample. Then a non-matching vpin is randomly selected. This pair constitutes a *negative* sample. Equal number of positive and negative sample pairs (v_1, v_2) are generated. For each pair, the parameters in Table 2.2 are computed.

Table 2.2 Parameters computed during sample generation [15]

Sl.	Parameter	Definition
1	DiffPinX	$|px_1 - px_2|$
2	DiffPinY	$|py_1 - py_2|$
3	ManhattanPin	$|px_1 - px_2| + |py_1 - py_2|$
4	DiffVpinX	$|vx_1 - vx_2|$
5	DiffVpinY	$|vy_1 - vy_2|$
6	ManhattanVpin	$|vx_1 - vx_2| + |vy_1 - vy_2|$
7	TotalWirelength	$W_1 + W_2$
8	TotalArea	$A_{in1} + A_{in2} + A_{out1} + A_{out2}$
9	DiffArea	$(A_{out1} + A_{out2}) - (A_{in1} + A_{in2})$
10	PlacementCongestion	$PC_1 + PC_2$
11	RoutingCongestion	$RC_1 + RC_2$

3. *Training:* A Bootstrap aggregation (Bagging) method [16] is used for classification. Bagging method combines the outputs of a group of base classifiers. Reduced error *pruning* tree [17], implemented as REPTree in Weka [18], is used as the base classifier. Leave-one-out method is used for cross-validation. Among the available designs, to test each design, the remaining designs are used to train the classifier. This ensures the separation of testing and *training* data sets.

4. *Testing:* For testing, the set of all possible vpin pairs in a given test design (with the same splitting layer for which the classifier was trained) are considered for evaluation. The classifier is used to predict if the pair is a match based on a yes/no answer. The "yes" matches for each vpin constitute its List of Candidate (LoC) matching pins. Two quality metrics for the classifier are defined: (1) Size of LoC, smaller size being better, and (2) Classification Accuracy (CA), defined as the inclusion of the correct match of a vpin in its LoC.

2.6.2 Further Enhancements

Scalability When splitting is done at lower layers, the number of vpins increases dramatically. This, in turn, leads to longer training and testing times. The classification accuracy drops due to "useless" negative samples that were selected at random regardless of their distance from the vpin. In addition, the size of the LoC for a candidate vpin increases indicating uncertainty in the recovery of the matching pair. To solve these scalability issues, a small neighborhood is defined around each vpin. All samples for training and testing are drawn only from the neighborhood for each vpin. To define the neighborhood, the value of ManhattanVpin is determined for each positive sample in the training data set. The neighborhood size is defined such that p% of all vpin pairs are in the neighborhood. p is nominally set to 90%. Decreasing p accelerates training and testing at the expense of accuracy.

Two-Level Pruning While improving accuracy, the use of the neighborhood as discussed leads to the inclusion of several negative samples in the LoCs. To further reduce the LoC size while retaining accuracy, Level-2 pruning is introduced. The idea is to train a Level-2 classifier only for those vpins in the Level-1 LoC with the goal of distinguishing the correct vpin from the incorrect ones. Following the use of Algorithm 3 for training the Level-1 classifier, a Level-1 LoC(v) for each vpin v in the training set is generated using the Level-1 model. Then, for each v, a non-matching vpin is selected at random from LoC(v). These "high-quality" negative samples along with all the positive samples are used to train a Level-2 classifier. During testing, given a target design and a target vpin, first the Level-1 classifier is used to generate the LoC. This LoC is then pruned using the Level-2 model which yields the final LoC.

Controlling the LoC Size Algorithm 3 uses a binary classifier resulting in a single value of accuracy for each LoC. However, when comparing models, it is sometimes useful to control the degree of accuracy by varying the size of LoC. Zeng et al.

proposed a simple method for this. The Bagging meta classification scheme used employs *soft voting* to combine the results from individual REPTree classifiers. Each classifier outputs a probability that the sample (v_1, v_2) is positive. The combiner computes the average probability $p(v_1, v_2)$. The sample is marked positive if $p(v_1, v_2) \geq t$, where t is a threshold set to 0.5 by default. By varying t, LoCs of different sizes, yielding different accuracy results, can be produced.

Special Case of Splitting In CMOS VLSI designs, it is common to route all wires in the same layer parallel to each other. In this case, if the splitting occurs at the highest via layer, that is, the BEOL consists only of the top metal layer, either DiffVpinX or DiffVpinY can be set to zero when generating training samples and ignoring ineligible pairs during testing.

Proximity Attack The proximity attack matches each target vpin v in the target design with the candidate pin $v' \in LoC(v)$ such that the distance $d(v, v')$ is minimized. $p(v, v')$ is used to break ties. The attack is successful only if (v, v') are actually connected. Attack success varies with the size of LoC which is a function of t. However, the same threshold value is unlikely to yield successful attack result for every pin; LoC could be empty for some pins and quite long for others. An LoC value for the attack is determined as follows: From the training set of designs, 80% of vpins are randomly selected and used to train a model as discussed. The remaining 20% of vpins in each design are used for validation. To account for the different numbers of vpins in different benchmarks, the validation is performed for different values of the *LoC fraction* defined as the ratio of LoC size to the number of vpins in the benchmark. The LoC fraction which results in the best attack success rate in the validation process averaged over all the $|\mathcal{D}|$ training designs should be used in the actual attack.

2.6.3   Discussion

ML models are highly sensitive to the quality of data used, requiring extensive experimental studies and training. Zeng et al. have extensively studied the performance of the ML approach to analyze how secure SM could be in the face of an ML attack and reported detailed experimental results [14, 15]. They have used the large ISPD-2011 benchmarks [19] which consist of up to nine metal layers.

In their experiments, the proximity attack using the trained classifier consistently produced better results than those reported by Magaña et al. [9], discussed in Sect. 2.4. For the same classification accuracy, the LoC size was less than 3% of the size reported in [9]. Conversely, for the same LoC size, the accuracy was close to 100%, compared to about 43% in [9]. Note that, the classifier does not directly recover the BEOL net. Instead, it narrows down the possible vpin matches to the LoCs. Attackers should use other design knowledge to find exact matches. Classifier performance can be quite sensitive to noise. Even a small amount of noise (1%–2%)

added to the y-coordinates of vpins can result in sufficient loss of accuracy resulting in a reduction of attack success rate by up to 81% at layer 6 and up to 55% at layer 4.

Experiments also show that the relative importance of features in decreasing order of importance is as follows: locations of vpins, distance between vpins, locations of pins in the placement layer, distance between pins, and diffArea. In general, routing has more impact than placement on security.

Zeng et al.'s studies demonstrated the potential of ML methods to attack SM. To further improve the attack correctness, use of other classifiers, for example, neural networks, and use of other feature sets, for example, image data, can be explored. Li et al. [20] have recently proposed a deep learning approach using both vector and image data to attack split circuits.

2.7 Simulated Annealing (SA) Based Trojan Insertion Attack

So far in this chapter, we have discussed design constraint based reverse engineering attacks whose goal is to reconstruct the entire circuit by recovering the missing BEOL signals. The remaining attacks in this chapter are concerted with *trojan insertion*. Their goal is to uniquely identify locations in the FEOL circuit that match given structural features even in the absence of the BEOL wires. The attacker is interested in inserting a trojan at the matching location. These attacks are also called the *layout recognition* attacks since the goal is to correctly locate a portion of the layout (or subcircuit inferred from the layout) as a correct match to a given circuit.

We begin with the layout recognition attack proposed by Chen et al. [21] based on repeated use of the *simulated annealing* (SA) search. SA is a well-known probabilistic search heuristic used for combinatorial optimization [22, 23].

2.7.1 Mapping Problem Formulation

The problem of matching a given netlist to a given layout is formulated as follows. Let V_n be the set of gates in the netlist. Let V_l be the set of gates in the FEOL layout. Let $m_c : V_n \to V_l$ be the correct *mapping* between V_n and V_l. m_c is a bijective mapping. Let $m : V_n \to V_l$ be the mapping recovered by the attacker whose goal is to ensure that m is as close as possible to m_c without, of course, knowing m_c.

Figure 2.6a shows an example netlist. Figure 2.6b shows a layout for this netlist. Assuming that all the nets are delegated to BEOL, Fig. 2.6c shows the attacker's view of the FEOL layout. In this example, $V_n = \{g1, g2, g3, g4, g5\}$ and $V_l = \{a, b, c, d, e\}$. The correct bijective mapping is $m_c(g1) = a, m_c(g2) = b, m_c(g3) = c, m_c(g4) = d, m_c(g5) = e$.

The attacker does not have to uniquely identify the location of a target gate $v \in V_n$. If v is the target for a HT implantation, the attacker would simply insert a trojan at multiple locations each of which is suspected to map to v. In order to admit the

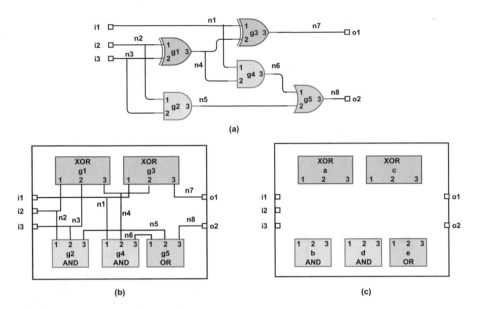

Fig. 2.6 Mapping from netlist to layout (based on [21]). (**a**) Example netlist, (**b**) Layout for the netlist, (**c**) Attackers view of the FEOL

uncertainty of the mapping, m is defined as $m : V_n \rightarrow \mathcal{P}(V_l)$ where $\mathcal{P}$ denotes the powerset of a set. For node $v \in V_n$, $m(v)$ is the subset of nodes in V_l that are potential candidates for the mapping of v. For convenience, m_c is similarly defined with the restriction that for all nodes $v \in V_n$, $|m_c(v)| = 1$. For the target node v, the attacker is hoping that $m(v) = m_c(v)$ or, at least, $m_c(v) \subseteq m(v)$ and $|m(v)|$ is as small as possible in order to reduce the cost of HT implantation while improving accuracy.

2.7.2 SA Based Mapping

SA based mapping proposed by Chen et al. [21] makes use of two observations: (1) For a netlist gate to match a layout gate, the type of both gates should be the same. (2) Layout synthesis tools generally estimate the wire length of a net using the half-perimeter method and attempt to minimize the total wire length [24].

Each gate has a type such as NAND, NOR, OAI, etc. Gate type can be easily identified in the layout since the complete layout for each cell is available in FEOL. For a netlist gate to be mapped to a layout gate, their types have to match. The mapping of the netlist input/output terminals to layout input/output terminals is assumed to be known to the attacker.

Algorithm 4: SA based layout recognition attack [21]

Input: Netlist Gates V_n, Layout Nodes V_l, Number of SA Attempts N, Pruning Factor α
Output: Mapping $m : V_n \rightarrow \mathcal{P}(V_l)$
    `// Mapping`
1 **for** $i = 1$ *to* N **do**
2   Generate a random bijective mapping $m_i : V_n \rightarrow V_l$ such that, for each $v \in V_n$, the
     type of $m_i(v)$ is the same as the type of v
3      $m_i = $ Simulated_Annealing(V_n, V_l, m_i);
4 **end**
5 **for** *each* $v \in V_n$ **do**
6      $m(v) = \bigcup_{i=1,N}\{m_i(v)\}$;
7 **end**
    `// Pruning`
8 **for** *each* $v \in V_n$ **do**
9   **for** *each* $u \in m(v)$ **do**
10          $n_{v,u} = |\{m_i | u \in m_i(v), i = 1, N\}|$;
11          $p_{v,u} = \frac{n_{v,u}}{N}$;
12   **if** $p_{v,u} < \frac{\alpha}{|S(v)|}$ **then**
13              $m(v) = m(v) - \{u\}$;
14   **end**
15   **end**
16 **end**
17 **return** m;

For the example in Fig. 2.6, initially $m(g1) = \{a, c\}$, $m(g2) = \{b, d\}$, $m(g3) = \{a, c\}$, $m(g4) = \{b, d\}$, $m(g5) = \{e\}$, $m(i1) = \{i1\}$, $m(i2) = \{i2\}$, $m(i3) = \{i3\}$, $m(o1) = \{o1\}$, and $m(o2) = \{o2\}$.

Algorithm 4 shows the SA based layout recognition attack. Initially, a number of viable bijective mappings between V_n and V_l are generated by repeatedly using an SA algorithm (lined 1–4). Each SA run starts with a randomly generated bijective mapping between V_n and V_l such that the gate types match (line 2). Then this mapping is optimized using the SA with the total *half-perimeter wire length* (HPWL) as the cost function. Thus, the final mapping produced is a heuristic estimate of m_c where the layout was likely produced using a similar wire length minimizing place and route system. In order to increase the chances of the actual mapping to be covered, the SA is repeated some N times and results are combined (lines 5–7) to obtain the final list of candidates in V_l to be mapped to each $v \in V_n$.

The next step is to prune the mapping using a simple heuristic. For each $v \in V_n$, let $S(v) \subseteq V_l$ denote the set of all gates in V_l whose type matches that of v. Then, $1/|S(v)|$ is the nominal probability of any of those gates being mapped to v. If any of those gates was mapped to v even fewer times by the SA runs than suggested by this probability, then that gate can be pruned from $m(v)$. A control parameter $\alpha \in [0, 2]$ is used to add additional flexibility. Higher value of α increases pruning and lower value reduces it.

For each netlist node v and each $u \in m(v)$, the algorithm determines the number of SA runs $n_{v,u}$ in which u appeared in the resulting bijective mappings. If the actual occurrence ratio $n_{v,u}/N$ is less than the nominal occurrence probability (modulated by α), then u is discarded from $m(v)$ (lines 8–16).

2.7.3   Evaluation Metrics

In order to evaluate the attack, Chen et al. [21] proposed two figures of merit:

1. *Effective Mapped Set Ratio (EMSR):* The mapping is effective if the actual mapped gate is contained in the candidate set for as many netlist gates as possible. EMSR is the percentage of netlist gates for which this happens.

$$\text{EMSR} = \frac{\sum_{v \in V_n} |m(v) \cap m_c(v)|}{|V_n|} \tag{2.7}$$

2. *Average Mapped Set Pruning Ratio (AMSPR):* Without the attack algorithm, each netlist node v has each layout node of the same type as a potential candidate for mapping. Thus, the alternative candidates for v are $S(v)$. The SA attack essentially prunes this uncertainty to $m(v) \subseteq S(v)$. The *Pruning Ratio* for v is defined as the normalized difference between the sizes of $S(v)$ and $m(v)$ provided $m(v)$ contains the actual mapped node of v per $m_c(v)$; otherwise, the pruning ratio for v is 0. AMSPR is the average pruning ratio over all the netlist nodes and is a measure of the effectiveness of the pruning process.

$$\text{AMSPR} = \frac{1}{|V_n|} \cdot \sum_{v \in V_n} \left(\frac{|S(v)| - |m(v)|}{|S(v)|} \right) \cdot |m(v) \cap m_c(v)| \tag{2.8}$$

2.7.4   Discussion

Chen et al. [21] evaluated the SA based mapping followed by probability based pruning using the ISCAS-85 benchmarks. The best value for N, the number of mappings found, was experimentally determined to be 2 × Maximum number of same-type gates. As α increases, EMSR decreases and AMSPR increases. α value of 0.9 was experimentally determined to yield good tradeoff between EMSR and ASMPR. At this value of α, the average EMSR and AMSPR are 85% and 50% respectively. In addition, most mapped sets were pruned to a much smaller size compared to their pre-attack sizes while still keeping them effective, i.e. each pruned set still includes the actual mapped gate.

Attack using repeated SA runs could be extremely time consuming. In addition, aggressive optimization of total wire length can mislead the attack since layout synthesis tools prioritize the net routing using considerations such as critical path timing and available slacks. The information about the net priority order is not available to this attack.

2.8   Proximity Based Mapping and Net Based Pruning

Algorithm 4 has two main steps: mapping and pruning. Mapping was achieved using an SA based approach and pruning was performed using a probability based heuristic. Yang et al. [25] proposed alternative algorithms for both steps. We will describe these alternatives in this section.

2.8.1   Proximity Based Mapping

Instead of optimizing the total wire length while constructing the mapped sets, the proximity based mapping algorithm uses the pin proximity information to map netlist gates to layout gates. The algorithm uses cell types, pin locations, and pin proximity information to mount the attack.

Each netlist gate has a fixed number of *pins*. The corresponding standard cell in the layout has fixed pin positions relative to each other and relative to the center of the cell. Pins are connected by *nets*. Each net connects a set of pins belonging to different cells. Two pins belonging to two different cells are said to be of the same *type* provided their gate types are the same and their relative positions in the respective cells are the same. The IO *terminals* of the design are also represented as gates and have one pin each.

In Fig. 2.7a, consider net $n5$ which has 2 terminals, $g2.3$ and $g5.2$. Assume that $g2$ is already mapped to cell b in the layout which means that $g2.3$ is mapped to $b.3$. Simple proximity attack would consider mapping $g5.2$ to $d.1$ since $d.1$ is the closest input pin to $b.3$. This would, of course, be incorrect. The proposed mapping method, which considers types of the gates while mapping the ports, does not consider $d.1$ to be an appropriate match to $g5.2$ since their (gate) types do not match. The only pin that matches pin $g5.2$ is $e.2$. In addition, once a gate is mapped, all its pins can be mapped. For example, if $g5.2$ is mapped to $e.2$, it implies that $g5$ is mapped to e and $g5.1$ and $g5.3$ are mapped to $e.1$ and $e.3$ respectively.

Algorithm 5 shows the proximity based mapping algorithm. The algorithm constructs two mappings: $m_g : V_n \rightarrow V_l$ and $m_p : P_n \rightarrow P_l$, where V_n and V_l are the sets of netlist gates and layout cells respectively (both sets include the IO terminals) and P_n and P_l are the sets of netlist pins and layout pins respectively. The mapping of the netlist IO terminals to the layout IO terminals is assumed to be known to the attacker. Once a gate is mapped, its pins can also be mapped. Similarly, once a pin of a gate is mapped, the gate itself can be mapped.

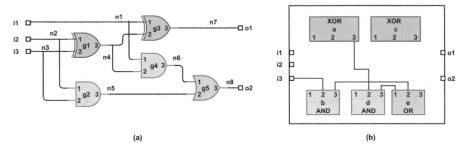

(a)   (b)

Fig. 2.7 Proximity based mapping (based on [25]). (**a**) Example netlist, (**b**) Mapping sequence

---

Algorithm 5: Proximity based netlist to layout mapping [25]

Input: Netlist Gates V_n, Netlist Pins P_n, Layout Nodes V_l, Layout Pins P_l
Output: Mappings $m_g : V_n \rightarrow V_l$, $m_p : P_n \rightarrow P_l$
1 Map the terminal nodes in V_n to terminal nodes in V_l;
2 Map the terminal pins in P_n to terminal pins in P_l;
3 g = a randomly selected terminal in V_n;
4 **while** *there are unmapped gates or pins* **do**
5   **if** *some non-terminal gates in V_n are mapped and all pins on every net connected to g
    are mapped* **then**
6   g = a randomly selected non-terminal gate in V_n such that g is mapped and has at
        least one net with unmapped pins;
7   **end**
8   p = a randomly selected pin on g such that p is on a net n with unmapped pins;
9   p_n = a randomly selected unmapped pin on n;
10   p_l = nearest pin to $m_p(p)$ in the layout such that p_l is of the same type as p_n;
11      $m_p(p_n) = p_l$;
12   g_n = netlist gate in V_n with which p_n is associated;
13   g_l = layout cell in V_l with which p_l is associated;
14      $m_g(g_n) = g_l$;
15   **for** *each unmapped pin i on g_n* **do**
16          $m_p(i) = j$ such that j is the unmapped pin corresponding to i on g_l;
17   **end**
18      $g = g_n$;
19 **end**
20 **return** m_g, m_p;

The algorithm begins by initializing the mappings for the terminal nodes and pins (lines 1–2). In each iteration of the while loop (lines 4–19), the algorithm selects a mapped gate $g \in V_n$ and a mapped pin p of g such that p is on a net n with one or more unmapped pins (line 8). Then, a randomly selected unmapped pin p_n on this net will be selected and mapped to a pin $p_l \in P_l$. p_l will be selected such that it is the nearest pin to $m_p(p)$ in the layout and is of the same type as p_n (lines 9–11). The gate of p_n will be mapped to the gate of p_l in the layout (lines 12–14). Once the gate is mapped, then its pins are also appropriately mapped (line 15–18). The algorithm then continues the same process from the newly mapped gate (line 18). If the newly mapped gate has no nets with unmapped pins, then another appropriate gate is selected at random (lines 5–7). For the first pass of the while loop, a terminal node is selected as the seed at random since the non-terminal gates are yet to be mapped (line 3).

Considering Fig. 2.7, assume that $i3$ is the initial terminal selected randomly (line 3). Its only pin is also named $i3$. It is connected to net $n3$ which has two unmapped pins, $g1.2$ and $g2.2$ (line 8). The algorithm randomly selects one of them, say, $g2.2$, as the next pin for mapping (line 9). In the layout, the pin closest to $i3$ and matches $g2.2$ is $b.2$. This is selected and $g2.2$ is mapped to $b.2$ (lines 10–11). Consequently, $g2$ is mapped to b (lines 12–14) and pins $g2.1$ and $g2.3$ are mapped to $b.1$ and $b.3$ respectively (lines 15–17). The next gate selected is $g2$. Any pin on $g2$ which is on a net which has unmapped pins can be selected next and the process is repeated. The red lines in Fig. 2.7b show the next few steps in the mapping sequence as follows: $g2.3$ selected; $g5.2$ selected; only matching port is $e.2$, hence, $m_p(g5.2) = e.2$; $m_g(g5) = e$; $m_p(g5.1) = e.1$; $m_p(g5.3) = e.3$; $g5.1$ selected; $g4.3$ selected; nearest matching port is $d.3$, hence, $m_p(g4.3) = d.3$; $m_g(g4) = d$; $m_p(g4.1) = d.1$; $m_p(g4.2) = d.2$; $g4.2$ selected; $g1.3$ selected; nearest matching port is $a.3$, hence, $m_p(g1.3) = a.3$; $m_g(g1) = a$; $m_p(g1.1) = a.1$; $m_p(g1.2) = a.2$. This process continues until all the gates and the ports are mapped.

2.8.2 Net Based Pruning

The mapping results produced by Algorithm 5 are highly dependent on the random choices made during the mapping sequence. For example, Fig. 2.8 shows two possible mappings (indicated in blue and pink) to the same portion in the layout depending upon whether $i1$ or $i3$ is chosen initially and the choices of nets and pins in the first few steps.

Therefore, different mapping solutions can be obtained by multiple runs of the algorithm. Suppose the algorithm is executed N times and $m_{g_i}, i = 1, N$ mapping

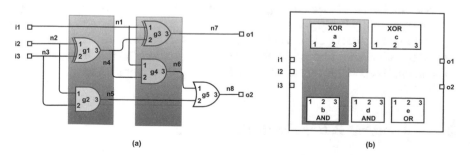

(a)   (b)

Fig. 2.8 Alternative mappings (based on [25]). (**a**) Two subcircuits, (**b**) Corresponding portion of the layout

results are obtained. These results are pruned as follows: for each mapping solution m_{g_i}, for each net n in the netlist, compute its half-perimeter wire length (HPWL), denoted by $L_i(n)$, $i \in [1, N]$. Retain the mappings of the gates connected by the net in every solution $i \in [1, N]$ in which the following inequality is satisfied:

$$L_i(n) \leq \underset{i=1,N}{Min}(L_i(n)) + \beta * (\underset{i=1,N}{Max}(L_i(n)) - \underset{i=1,N}{Min}(L_i(n))) \qquad (2.9)$$

If this inequality is not satisfied, then the mappings of the gates connected by net n in those solutions are further pruned by the probability based pruning method discussed in Algorithm 4. Finally, all the mappings remaining after the pruning are considered by the attacker as potential sites for trojan insertion for the target gates. $\beta \in [0, 1]$ is a control parameter which determines the number of mappings pruned. As the β value is increased, solutions with longer nets tend to be retained as acceptable mappings.

2.8.3 Discussion

Yang et al. [25] evaluated and compared the SA based mapping and the proximity based mapping methods using the ISCAS-85 and ITC-99 benchmarks. The best value for N, the number of mappings found, was determined to be $2 \times$ Maximum number of same-type gates. After repeating N runs and merging the results, with further pruning, the SA method was observed to be superior with average EMSR and AMSPR values of 95.79% and 40.13% respectively. These values for the proximity based mapping method were 82.82% and 30.89% respectively. This implies that the global total wire length metric was better than the local proximity metric. This is not surprising in view of the observations made in Sect. 2.4. However, SA takes significantly longer time than the proximity method.

As β decreases, AMSPR increases and EMSR decreases. $\beta = 0.3$ was found to result in good tradeoff between AMSPR and EMSR. The two pruning methods were compared with $\alpha = 0.9$ and $\beta = 0.3$ with the SA as the mapping method. Results indicated that the net based pruning method improves EMSR by 5.59% on average over the probability based pruning.

Note that the attacks discussed in this section and the previous section return a pruned set of mappings but do not necessarily an exact mapping even if one exists. In addition, it is possible to omit the actual mapping from a pruned set.

2.9 Structural Pattern Matching Attack

The attacks discussed so far have exploited design constraints based on proximity and other information extracted from the FEOL layout. Adding logical reasoning to the attack process can help speed up the attack while avoiding incorrect mappings. Xu et al. [26] proposed a layout recognition attack based on exploiting proximity information along with simple logical reasoning. The attack uses a mapping method inspired by a technology mapping algorithm [27].

2.9.1   Pattern Tables

To facilitate *pattern matching* between the netlist and the layout, the information in the netlist is arranged in the form of *pattern tables*. Each logic gate is assigned an index. The connectivity relationships among the gates are represented by tables, one for each gate type. In each table, the input columns are divided into groups such that the inputs in each group denote commutative inputs whose order can be interchanged. Given a netlist, a pattern table representation can be generated by traversal in the topological order. When matching an entire circuit, the mapping of input and output terminals is assumed to be known. In this case, each terminal has a unique index. However, when matching a subcircuit, the input indices are all set to 0 indicating that their mapping can be treated as a don't care.

Figure 2.9 shows a netlist and the corresponding pattern tables. Each gate type has a table. There are two instances each of the INV and NOR2 gates and one instance each of the others. Input to NOR2, NAND2, XOR2, and AND2 are arranged as a single group since they are all commutative. Inputs to IAND2 are arranged in two separate groups since they are not interchangeable. Each table has one row for each instance of the gate.

2.9.2   Pattern Matching

To match a subcircuit, from the FEOL layout, all the gates are extracted and a partial netlist is initialized. Initially, indices of all layout gates are set to 0. For each layout gate g_l, a set of candidate gates in the netlist with which g_l can be mapped is maintained. Initially, for g_l, all the netlist gates whose type is the same as that of g_l form the candidate set. For example, Fig. 2.10 shows the layout view of the circuit where solid lines indicate the FEOL nets and the dashed lines indicate the BEOL nets. Cells A', D', E', and H' correspond to unique instances of gates in the netlist. Hence, their mapping can be readily inferred. For the remaining cells, B', C', G', and F', there are multiple candidate gates. These candidate sets are pruned in three steps:

1. *Structural matching and logical reasoning:* The mapping is pruned using the pattern tables and logical reasoning about the connections. Once a match is found, the index of the layout gate is changed to that of the matched netlist gate. In Fig. 2.10, the pattern (NOR2,0,0) of B' matches 2(NOR2,0,0) in the netlist.

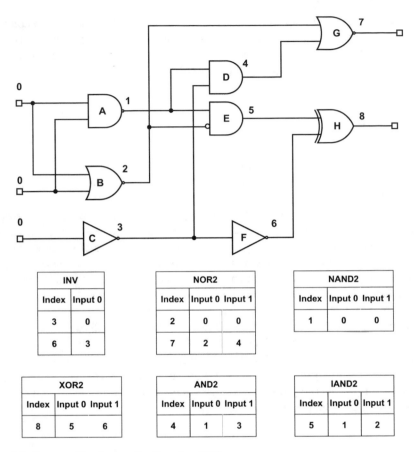

Fig. 2.9 Pattern tables for a netlist (based on [26])

Hence, the index of B' is changed to 2. Similarly, (INV, 0) matches 3(INV,0). Hence, index of C' is changed to 3. Candidate list for G' can now be pruned to {7} since 2 has already been matched to B'. Similarly, index of F' can now be set to 6. An input of H' was initially uncertain, "5/6." However, since the other input is set to 5, this input can only be 6. For gate D', the two inputs have two candidates. But, since the cell is commutative, these can be arbitrarily fixed. Logical reasoning with the help of the pattern tables continues this way until no further inferences can be drawn.

2. *Layout hints:* In the next step, further pruning is performed using hints from the layout discussed in Sect. 2.5: (1) load capacitance constraint, (2) timing constraint, and (3) direction of dangling wires. After this pruning, the mapping that yields the minimum BEOL wire length is selected.

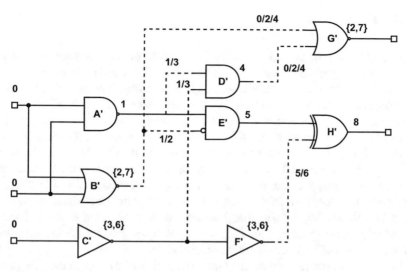

Fig. 2.10 Structural matching using pattern tables (based on [26])

Algorithm 6: Structural pattern matching attack [26]

Input: Netlist Gates V_n, Layout Gates V_l
Output: Mappings $m : V_n \rightarrow \mathcal{P}(V_l)$
1 Assign unique indices to V_n and prepare pattern tables.
2 for *each $v \in V_l$* **do**
3 $\quad m(v) = \{$index of $g \in V_n$ | gate type of v matches that of $g\}$
4 end
5 $m = $ structural_matching_and_logical_reasoning_based_pruning(m);
6 $m = $ layout_hints_based_pruning(m);
7 $m = $ subcircuit_propagation_based_pruning(m);
8 return m;

---

3. *Subcircuit propagation:* If some gates still have multiple candidate indices, then they are propagated toward the output terminals. When a candidate pattern is combined with that of a fanout gate and the combined pattern cannot be matched with the patterns in the pattern tables, the candidate can be pruned. In general, subcircuit topologies are directed acyclic graphs. To reduce the probability of illegal matches due to inconsistencies of propagation along multiple paths, tree subcircuits are processed first.

The matching process is summarized as Algorithm 6. To match an entire circuit, the circuit is divided into subcircuits and each subcircuit is matched separately. Processing proceeds from inputs to outputs in topological order. When some gates are recognized, their fanout gates are processed next. Matching terminates when layout gates cannot be further recognized.

2.9.3 Discussion

Xu et al. [26] used the ISCAS-85 and ITC-99 benchmarks to compare the structural pattern matching method with the satisfiability (SAT) based bijective mapping method [28] (to be discussed in Sect. 4.2). Their experiments showed that the pattern matching method achieves the same level of *matching ratio* (proportion of netlist gates correctly identified in the layout) as the SAT based bijective mapping and does so in a much shorter time. When the k-security defense [28] (Sect. 1.12) for k values of 2, 3, or 4, is incorporated to defend 5–10% of the cells, the pattern matching method achieved a higher matching ratio than the expected success rate. Using the k-security method, the defended cells have a *probability* of less than 1/k to be correctly identified. In their experiments, the pattern matching method was able to beat these odds and performed better than the SAT based method. However, the pattern matching method took longer in some cases since the netlist was divided into subcircuits and each subcircuit was matched individually.

The pattern matching method may not be scalable to large circuits. The interaction between structural matching and pruning based on layout hints requires further study.

2.10 Summary

Table 2.3 shows a summary of the design constraint based attacks discussed in this chapter. In general, success of these attacks depends on the exploitable information an attacker can obtain from the FEOL layout, the attacker's knowledge of the CAD methods used, and the sequence in which the attack method utilized the available information. As we will see in the next chapter, effective defense methods have been developed to thwart these design constraint based attacks.

Table 2.3 Summary of design constraint based attacks

Sl.	Attack	Year	Type	Benchmarks	Metrics
1	Proximity attack	2013	RE	ISCAS-85	AC
2	Extended proximities	2016	RE	ISPD-11	E[SA], E[LS], E[LS/SA]
3	Network flow attack	2016	RE	ISCAS-85, ITC-99	AC, OER
4	Machine learning attack	2018	RE	ISPD-11	LoC, CA
5	Simulated annealing attack	2016, 2020	TI	ISCAS-85, ITC-99	EMSR, AMSPR
6	Proximity based mapping and net based pruning	2020	TI	ISCAS-85, ITC-99	EMSR, AMSPR
7	Structural pattern matching attack	2019	TI	ISCAS-85, ITC-99	Matching ratio, k-security

RE, Reverse engineering attack; TI, Trojan insertion attack

References

1. M. Jagasivamani, P. Gadfort, M. Sika, M. Bajura, M. Fritze, Split-fabrication obfuscation: metrics and techniques, in *Proceedings of the 2014 IEEE International Symposium on Hardware-Oriented Security and Trust, HOST 2014* (2014), pp. 7–12
2. J. Rajendran, O. Sinanoglu, R. Karri, Is split manufacturing secure? in *Proceedings – Design, Automation and Test in Europe* (2013), pp. 1259–1264
3. Y. Xie, C. Bao, A. Srivastava, Security-aware design flow for 2.5D IC technology, in *TrustED 2015 – Proceedings of the 5th International Workshop on Trustworthy Embedded Devices, co-located with CCS 2015* (2015), pp. 31–38
4. J. Rajendran, H. Zhang, C. Zhang, G.S. Rose, Y. Pino, O. Sinanoglu, R. Karri, Fault analysis-based logic encryption. IEEE Trans. Comput. **64**(2), 410–424 (2015)
5. J. Rajendran, Y. Pino, O. Sinanoglu, R. Karri, Logic encryption: a fault analysis perspective, in *Proceedings – Design, Automation and Test in Europe* (2012), pp. 953–958
6. L. Lavagno, I.L. Markov, G.E. Martin, L.K. Scheffer, *Electronic Design Automation for IC System Design, Verification, and Testing* (CRC Press, Boca Raton, 2017)
7. L. Scheffer, L. Lavagno, G. Martin, *EDA for IC Implementation, Circuit Design, and Process Technology* (CRC Press, Boca Raton, 2006)
8. J. Magaña, D. Shi, A. Davoodi, Are proximity attacks a threat to the security of split manufacturing of integrated circuits? in *IEEE/ACM International Conference on Computer-Aided Design, Digest of Technical Papers, ICCAD*, vol. 07 (2016), pp. 1–7
9. J. Magana, D. Shi, J. Melchert, A. Davoodi, Are proximity attacks a threat to the security of split manufacturing of integrated circuits? IEEE Trans. Very Large Scale Integ. (VLSI) Syst. **25**(12), 3406–3419 (2017)
10. Y. Wang, P. Chen, J. Hu, G. Li, J. Rajendran, The cat and mouse in split manufacturing. IEEE Trans. Very Large Scale Integ. (VLSI) Syst. **26**(5), 805–817 (2018)
11. Y. Wang, P. Chen, J. Hu, J.J. Rajendran, The cat and mouse in split manufacturing, in *Proceedings – Design Automation Conference*, vol. 05 (2016), pp. 1–6
12. D.P. Williamson, *Network Flow Algorithms* (Cambridge University Press, Cambridge, 2019)
13. F. Lang, W. Yujie, H. Jiang, R. Jeyavijayan (JV), Network Flow Attack Software, https://github.com/seth-tamu/network_flow_attack (also at https://cadforassurance.org/tools/ip-ic-protection/network-flow-attack-for-split-manufacturing/)
14. B. Zhang, J.C. Magana, A. Davoodi, J.C. Magañá, A. Davoodi, Analysis of security of split manufacturing using machine learning, in *Proceedings – Design Automation Conference*, vol. Part F1377 (2018), pp. 1–6
15. W. Zeng, B. Zhang, A. Davoodi, Analysis of security of split manufacturing using machine learning. IEEE Trans. Very Large Scale Integ. (VLSI) Syst. **27**(12), 2767–2780 (2019)
16. S.B. Kotsiantis, Bagging and boosting variants for handling classifications problems: a survey. Knowl. Eng. Rev. **29**(1), 78–100 (2014)
17. W.N.H.W. Mohamed, M.N.M. Salleh, A.H. Omar, A comparative study of reduced error pruning method in decision tree algorithms, in *Proceedings – 2012 IEEE International Conference on Control System, Computing and Engineering, ICCSCE 2012* (2012), pp. 392–397
18. I.H. Witten, E. Frank, M.A. Hall, C.J. Pal, *Data Mining: Practical Machine Learning Tools and Techniques*. Morgan Kaufmann Series in Data Management Systems (2016)
19. N. Viswanathan, C.J. Alpert, C. Sze, Z. Li, G.J. Nam, J.A. Roy, The ISPD-2011 routability-driven placement contest and benchmark suite, in *Proceedings of the International Symposium on Physical Design* (2011), pp. 141–146
20. H. Li, S. Patnaik, M. Ashraf, H. Yang, J. Knechtel, B. Yu, O. Sinanoglu, E.F. Young, Deep learning analysis for split manufactured layouts with routing perturbation, in *IEEE Transactions on Computer-Aided Design of Integrated Circuits and Systems* (2020), pp. 1–14

21. Z. Chen, P. Zhou, T.-Y. Ho, Y. Jin, How secure is split manufacturing in preventing hardware trojan? in *Proceedings of the IEEE Asian Hardware-Oriented Security and Trust (AsianHOST'16)* (2016), pp. 1–6
22. D. Bertsimas, J. Tsitsiklis, Simulated annealing. Stat. Sci. **8**(1), 10–15 (1993)
23. P. Siarry, *Metaheuristics* (Springer International Publishing, Berlin, 2016)
24. N.A. Sherwani, *Algorithms for VLSI Physical Design Automation* (Springer Science & Business Media, Berlin, 2012)
25. Y. Yang, Z. Chen, Y. Liu, T.Y. Ho, Y. Jin, P. Zhou, How secure is split manufacturing in preventing hardware trojan? ACM Trans. Des. Autom. Electron. Syst. **25**(2), 1–23 (2020)
26. W. Xu, L. Feng, J.J. Rajendran, J. Hu, Layout recognition attacks on split manufacturing, in *Proceedings of the Asia and South Pacific Design Automation Conference, ASP-DAC*, no. ii (2019), pp. 45–50
27. M. Zhao, S.S. Sapatnekar, A new structural pattern matching algorithm for technology mapping, in *Proceedings – Design Automation Conference* (2001), pp. 371–376
28. F. Imeson, A. Emtenan, S. Garg, M.V. Tripunitara, Securing computer hardware using 3D integrated circuit (IC) technology and split manufacturing for obfuscation, in *Proceedings of the 22nd USENIX Security Symposium* (USENIX Association, Berkeley, 2013), pp. 495–510

Chapter 3
Defenses Against Design Constraint Based Attacks

Abstract Design constraint based attacks discussed in the previous chapter exploit information which can inferred from the FEOL layout combined with the knowledge of commonly used design automation algorithms to recover the missing BEOL connections or to locate suitable locations for trojan insertion. In this chapter, we present several methods to thwart the design constraint based attacks. These defense methods aim to either increase the cost of the attack or confuse the BEOL signal recovery process mounted by the attack. We begin with a summary of the metrics which can be used to develop effective defense methods, general defense strategies, and the costs of defense. We discuss several defense methods including pin swapping, secure min-cut bipartitioning, secure multiway partitioning, placement perturbation, routing perturbation, concerted wire lifting, netlist clustering, artificial routing blockage insertion, and netlist randomization and summarize their effectiveness in defending against the design constraint based attacks, specifically, the proximity and the network flow attacks.

In the previous chapter, we have discussed several design constraint based attacks that exploit information extracted from the FEOL layout combined with the knowledge of how placement and routing algorithms work. In this chapter, we will discuss approaches to defend against such attacks by careful selection of the BEOL nets and by making selective changes to the netlist composition, cell binding, and placement and routing of the design. We first introduce metrics to guide the development of effective defense methods.

3.1 Defense Metrics

Defense methods make it harder or impossible for the attacker to recover the entire circuit from the knowledge of the FEOL layout alone. At the least, the attack time should significantly increase and attack correctness should decrease. Jagasivamani

© The Author(s), under exclusive license to Springer Nature Switzerland AG 2021
R. Vemuri, S. Chen, *Split Manufacturing of Integrated Circuits for Hardware Security and Trust*, https://doi.org/10.1007/978-3-030-73445-9_3

et al. [1] proposed several quantifiable metrics to evaluate the potential effectiveness of the defense methods based on the generic attack process discussed in Sect. 2.1.

1. *Neighbor Connectivity (NY):* This is a measure of the degree of connectivity among neighboring cells within a given distance, d. Given a rectangular placement grid on which cells are placed, let $N(c, d)$ be the number of cells within a distance d from a cell c, and let $M(c, d)$ be the number of cells among them to which c is connected. Then Neighbor Connectivity within a neighborhood d is defined by

$$NY(d) = \frac{\sum_c M(c, d)}{\sum_c N(c, d)} \tag{3.1}$$

   Similar definitions can be used to define neighborhood connectivities among pins, vpins, etc. A low NY value suggests a low degree of connectivity among neighboring cells (or pins) and generally makes it harder for an attacker to infer missing connections.
2. *Composition Index (CI):* This is a metric to determine if the design leaks any information due to the mix of cells used. Certain type of cells, example EX-OR gates, may leak useful information. Let $T(c)$ denote the type of a cell, and let the set of cells in a design be denoted by C. Let H denote the set of all cell types that can potentially leak information. These are called hint, base, bias, or anchor cell types. In a type-balanced design, each type of bias cell will have the same proportion of instances among all the bias cell instances. Composition index for any base cell type $t \in H$ is defined by [1, 2]

$$CI(t, H) = \frac{|\{c|T(c) \in H\}|}{|C|} \cdot \left| \frac{|\{c|T(c) = t\}|}{|\{c|T(c) \in H\}|} - \frac{1}{|H|} \right| \tag{3.2}$$

   If a base cell type has proportionate representation among all occurrences of base cells, then its CI value will be close to zero.
   Composition index for the class of base cell types H is defined by

$$CI(H) = \sum_{t \in H} CI(t, H) \tag{3.3}$$

   CI close to zero indicates that all bias cell types are evenly represented in the design and larger values of CI indicated uneven or skewed distribution of the bias cell types.
3. *Entropy:* Entropy function is used to measure the level of disorder in the FEOL design based on the variety among the cells used in a design. For a cell type t, let $p_t = |\{c|T(c) = t\}|/|C|$ denote the probability that a cell is of type t in a design. Entropy of the design is computed by

$$E = -\sum_t p_t \cdot \log(p_t) \tag{3.4}$$

If all the cells are of the same type, the entropy is low and design leaks little information based on cell type identification.

While these metrics are not necessarily directly used to evaluate various defense methods, they do provide general directions to the development of effective defense methods.

3.2 General Defense Methods

Design methods to confuse the attacker and increase the attack time can be classified as follows [3]:

1. *Netlist Obfuscation:* Netlist obfuscation methods modify the original netlist during the design process so as to confuse the attacker and force the attack to recover an incorrect netlist. The goal of netlist obfuscation is to ensure that the reconstructed netlist has as little resemblance as possible to the original netlist. Attack correctness (AC) metric discussed in Sect. 2.2 can be used to quantify netlist obfuscation. In the case of HT insertion attacks, the goal of netlist obfuscation is to ensure that the attacker cannot uniquely identify the target location. In this case, the k-security metric discussed in Sect. 1.12 can be used to quantify netlist obfuscation.
2. *Layout Obfuscation:* Layout obfuscation methods modify the design layout such that the correct BEOL connectivity cannot be inferred by an attack. Layout can be modified during placement or routing or both. As before, the AC metric and the k-security metric can be used to quantify the effectiveness of layout obfuscation.
3. *Functional Obfuscation:* The goal of functional obfuscation is to ensure that the function of the attacker's extracted design is substantially different from the function of the original design. Hamming distance based metrics discussed in Sect. 2.2 can be used to quantify the level of functional obfuscation. An average HD of 50% is assumed to be ideal functional obfuscation. In practice, functional obfuscation is achieved through the structural or layout obfuscation methods.

Based on the metrics discussed in the previous section, Jagasivamani et al. [1] proposed several general methods for defending against various forms of attacks.

1. *Limited standard cell library:* This method limits the number of cells used in the design. For synthesized designs, this method limits the standard cells made available to the synthesis tool to a few basic types and removes all the complex cells that might provide functional clues to an attacker.
2. *Dummy cell insertion:* Dummy cells that are not used can be added to confuse the attacker. In particular, dummy cells similar to the hint cells can be added to reduce the CI value and avoid leaking information useful to the attacker.
3. *Using isomorphic cells:* This method replaces standard cells in a design with generic *programmable* cells capable of performing multiple functions. Universal functional cells such as multiplexors and look-up tables can be used in the FEOL

layer and their function can be configured in the BEOL layer. For example, a two-input isomorphic cell implemented as a look-up table (LUT) can be configured to realize any of the 16 two-input functions depending on the 4 configuration selector signals. The specific configuration to implement a function is done by connecting the selector signals to power strips in the BEOL layers. Alkabani et al. [4] and Masoud et al. [5] suggested algorithms to restructure designs based on isomorphic cells.

4. *Suboptimal placement:* Proximity based recovery of BEOL nets is based on the wire length minimal placement and routing commonly performed by layout synthesis tools. In order to mitigate proximity based attacks, cells can be moved apart. As this increases wire length and decreases performance, it should be applied to cells connected to non-critical nets.

3.3 Defense Cost

The cost of a defense method can be measured in terms of the following parameters:

1. *Fabrication Cost:* Since SM requires separate fabrication of the BEOL part and merging of the BEOL and FEOL parts, the overall fabrication cost is a concern. The BEOL manufacturing cost is assumed to be directly proportional to the number of nets that need to be routed through the BEOL layers. Some defense methods increase the number of BEOL nets to improve security thereby adversely influencing the BEOL fabrication cost.

2. *PPA Costs:* Defense methods can adversely impact performance, power, and area (PPA) of the design. Security benefit obtained should be weighed against these overheads. Selective application of defense methods to specific portions of the design, such as non-critical paths, may partly mitigate these overheads while providing some security enhancement. PPA overheads can also be measured indirectly by wire length estimates. The half-perimeter wire length (HPWL) is a widely used metric for estimating the wire lengths of multi-terminal nets.

3. *Design Time:* Incorporating a defense method during the design process usually increases the design time during either logic synthesis or physical synthesis or both. For the defense methods to be practically usable, their impact on the design time should not be excessive for large designs.

3.4 Pin Swapping

The proximity attack, discussed in Sect. 2.3, connects each unconnected input pin of a partition to the nearest unconnected output pin of the other partition. Rajendran et al. [6, 7] proposed the pin swapping method to confuse the attacker attempting to exploit the proximity information.

3.4.1 Objective

The objective of the pin swapping method is to swap the output pins of a partition such that if the attacker used the swapped pin to recover the BEOL signal, the functionality of the recovered circuit would be significantly different from the original circuit. The average Hamming Distance (HD) between the output vectors of the original circuit and the incorrectly recovered circuit when a sufficiently large number of inputs are applied is used to characterize their functional difference. The goal is to swap enough pins to achieve an average HD of 50%, which implies maximum deviation in functionality. Corrupting output bits by pin swapping is akin to fault detection where a fault on the swapped pins should be activated and propagated to the primary outputs [6, 8–10]. The *fault impact (FI) factor*, defined as follows, is used to determine the quality of swapping an output pin t with another output pin s of the same partition:

$$FI_{t,s} = \sum_{i=1}^{\substack{\text{\# test patterns}}} \text{\# corrupted output bits when } s \text{ and } t \text{ are swapped} \qquad (3.5)$$

3.4.2 Algorithm

Algorithm 7 shows the pin swapping method. Given the circuit and partition information, the algorithm returns pairs of output pins of the partitions to be swapped. The algorithm iteratively selects these pairs until the 50% HD criterion is satisfied or all partition output pins have been swapped (line 4). The average HD between the original circuit C and the modified circuit C_s (resulting from the swapped pins) is determined by comparing their outputs over a sufficiently large number of input vectors. In each iteration, the method builds a list of candidate swapping pins for each available output pin (line 6). For an available output pin t, another available output pin s is a potential candidate for swapping provided s and t belong to the same partition, and swapping s and t does not introduce a combinational cycle in the circuit (lines 19–26). The fault impact factor for each potential swapping pair is computed (lines 7–9), and the pair that has the maximum fault impact is selected (line 11). The selected pair is included in the list of pins to be swapped, both pins are removed from the list of available pins, and the partition and circuit structures are updated (lines 12–15).

3.4.3 Discussion

The pin swapping method was originally proposed as a defense against the proximity attack discussed in Sect. 2.3 [6]. Rajendran et al. showed that a small

---

Algorithm 7: Pin swapping [6]

    Input: Circuit C, Partitions P
    Output: Set of <Target Pin, Swapping Pin> Pairs L
1 $L = \emptyset$;
2 U = all partition output pins;
3 $C_s = C$;
4 **while** $(U \neq \emptyset)$ *or* $(HD(C, C_s) < 50\%)$ **do**
5   **for** *each output pin* $u \in U$ **do**
6 S_u = Candidate_Swapping_Pins(u, U);
7   **for** *each* $s \in S_u$ **do**
8 $FI_{u,s}$ = Fault_Impact(u, s, C_s);
9   **end**
10   **end**
11 $(t, s) = \{t \in U, s \in S_t \mid FI_{t,s} = \underset{i,j \in U}{MAX}\, FI_{i,j}\}$;
12 $L = L \cup \{<t, s>\}$;
13 $U = U - \{t, s\}$;
14 P = Update_Partition(P, t, s);
15 C_s = Update_Circuit(C_s, t, s);
16 **end**
17 **return** L;
18 ——
19 **Candidate_Swapping_Pins** *(t, U)*

    Input: Target Pin t, Unassigned Candidate Pins U
    Output: Set of Candidate Swapping Pins S
20 $S = \{s \in U \mid s \neq t, s$ *is in the same partition as* $t\}$;
21 **for** *each pin* $s \in S$ *where* $s \neq t$ **do**
22   **if** *swapping* t *and* s *causes a combinational loop in the circuit* **then**
23 $S = S - \{s\}$;
24   **end**
25 **end**
26 **return** S;

---

number of swaps (around 20 in most cases but no more than 72 in the worst case) were sufficient to satisfy the 50% HD criterion for the ISCAS-85 benchmarks. With these swaps, although the proximity attack was able to recover, on average, 87% of the connections correctly, the remaining 13% incorrect connections were sufficient to result in 42% HD. Without the defense, the proximity attack achieved 96% correct connections with a HD of about 10% on average (below 6% in most cases). This demonstrates the effectiveness of the pin swapping method—with relatively few swaps, the outputs of the recovered circuits are significantly corrupted.

    Similar to the proximity attack that it defends against, the pin swapping method continues to assume that the BEOL nets for SM were identified using min-cut partitioning followed by aggressive wire length minimal placement and routing. However, identification of the BEOL nets by traditional min-cut bipartitioning strategy is questionable.

3.5   Secure Min-Cut Bipartitioning and Placement

What is the right strategy to select the nets to be lifted? One strategy pursued by researchers is to weigh the nets based on their importance to security and select the ones that, if cut, would result in enhanced security. The cut nets would be delegated to the BEOL layers. Xie et al. [11] proposed secure minimal cutset partitioning and secure placement algorithms for enhancing the security of 2.5D ICs against proximity based attacks.

3.5.1   Objective

The design is assumed to be partitioned into two circuits, both of which are fabricated at an untrusted foundry. The cut nets connecting the two circuits are implemented on an interposer layer fabricated at a trusted foundry. The complete IC is assembled at a trusted 2.5D integration facility. While the standard min-cut algorithms can be used to minimize the cut nets for 2.5D fabrication, they usually compromise security. Consider the standard min-cut partition of a netlist in Fig. 3.1a. The functionality of the primary output PO_1 is exposed since its input

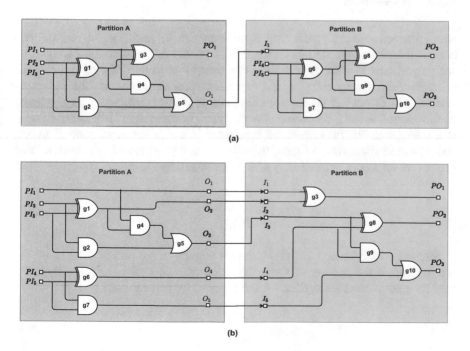

Fig. 3.1 Secure min-cut partition (based on [11]). (**a**) Normal min-cut partitioning, (**b**) Secure partitioning

cone is entirely within one FEOL partition. Primary inputs PI_4 and PI_5 cannot influence the cut signal. This reduces corruptibility of the output signals when the attacker infers the missing BEOL connections incorrectly. Output corruption is directly related to the activation and propagation of fault effects on the cut signals to the outputs as discussed in the previous section. *Secure partitioning* ensures that at least one path between every connected PI/PO pair is cut as illustrated in Fig. 3.1b. However, secure partitioning increases the number of cut nets. The goal of the secure min-cut partitioning algorithm is to identify the minimal set of nets to cut to bipartition the netlist while ensuring security and balancing the areas of the two partitions.

The security objective is to ensure that the proximity based attack can recover as few hidden nets as possible, and the incorrectly recovered netlist produces outputs that are maximally uncorrelated with those produced by the correct circuit. Formally,

$$\text{Minimize} \quad (|\text{ANHD} - 50\%| + AC) \tag{3.6}$$

where ANHD and AC are defined in Sect. 2.2.

3.5.2 Algorithm

Area-balanced min-cut partitioning algorithms can be used to generate the two circuits intended for 2.5D fabrication. However, the minimal cutset identified by such algorithms is unlikely to satisfy the security objective. Xie et al. proposed a modification incorporating a security-driven cost function into the Fiduccia–Mattheyses min-cut partitioning algorithm [12, 13]. This cost function itself is based on the *controllability* and *observability* measures introduced in [14]. Let w be a net in a netlist, and let i be one of the p primary inputs in the input cone of w. Let $c(w, i)$ denote the number of times the value on w flipped as i flipped when n input vectors are applied. Controllability of w is measured by

$$C(w) = \frac{1}{p} \sum_{i=1}^{p} \frac{c(w, i)}{n} \tag{3.7}$$

Similarly, let o be one of the q primary outputs in the output cone of w. Let $c(w, o)$ denote the number of times the value on o flipped as w flipped when n random input vectors are applied. Observability of w is measured by

$$O(w) = \frac{1}{q} \sum_{o=1}^{q} \frac{c(w, o)}{n} \tag{3.8}$$

The security metric $S(w)$ of net w is defined as

$$S(w) = C(w) + O(w) \tag{3.9}$$

The modified Fiduccia–Mattheyses (FM) algorithm is shown as Algorithm 8. The algorithm initially generates a bipartition of the netlist with both partitions having approximately the same area (line 1). In each pass (lines 3–11), the algorithm attempts to reduce the current cutset. A pass begins by unlocking all the cells and then iteratively selecting cells for tentative moves and entering them in a queue (lines 4–9). In each iteration, a cell c is considered for a move provided the security metric of the wire w driven by c is below a specified threshold $S_t \in [0, 2]$ and, if the cell were to move, the area imbalance defined as $AI = |A(N_1) - A(N_2)|/(A(N_1) + A(N_2))$, where $A(N)$ denotes the area of netlist N, will not exceed a specified threshold $A_t \in [0, 1]$. From among the considered cells, a cell c_i that yields maximum cutset reduction (even if such reduction is negative) is selected, tentatively moved to the other partition and locked. Once the tentative moves are queued, a subqueue is selected to maximize the cumulative gain G as long as G is positive (line 10). The moves in the subqueue are made permanent (line 11) and next pass begins.

Since the outcome of the FM algorithm is sensitive to the initial partition, the algorithm is executed several times and the solution with the least cutset size is selected. Algorithm 8 addresses the HD criterion.

To address the AC criterion, Xie et al. [11] proposed a secure placement method that uses the B*-tree [15] data structure to represent placement information. It uses the simulated annealing (SA) algorithm for 2.5D placement proposed in [16]. The method is as follows:

1. For each partition netlist generated from Algorithm 8, create a B*-tree to represent the placement of the gates and IO buffers in the netlist. The two B*-

---

Algorithm 8: Modified FM algorithm for secure partitioning [11, 13]

Input: Netlist N, Area Balance Threshold A_t, Security Threshold S_t
Output: Sub-Netlists N_1, N_2

1 (N_1, N_2) = Generate initial bipartition;
2 **repeat**
3   Unlock all cells;
4   **for** $i \in [1..|N|]$ **do**
5   Select a cell c_i with maximum cutset gain among all unlocked cells such that
           $S(w_i) < S_t$ and, if c_i were to move then $AI \le A_t$ after the move;
6   g_i = cutset gain of moving c_i;
7   Tentatively move c_i to the other partition;
8   Lock c_i;
9   **end**
10   Find k such that $G = \sum_{i=1}^{k} g_i$ is maximized;
11   Make moves of $c_1, \ldots c_k$ permanent and discard the other moves;
12 **until** $G \le 0$;
13 **return** N_1, N_2;

trees are used for simultaneous optimization of the placements within the two chips as follows.

2. Use an SA based placement algorithm to optimize the cost function $\alpha_1.Area + \alpha_2.L_{intra} + \alpha_3.L_{inter}$, where $Area$ is the normalized total area of the two placements, L_{intra} is the normalized total intra-chip wire length, L_{inter} is the normalized total inter-chip wire length, and the α's are user-specified weights. The following perturbation functions are used in the SA: (a) rotation of a gate or an I/O buffer, (b) moving a gate or an I/O buffer within the same B*-tree, and (c) swapping the locations of two gates or I/O buffers within the same B*-tree. The result of this SA is a compact placement.

3. A security-driven placement is now obtained by running the SA based placement algorithm on the placement obtained in the previous step. This run is the same as the one used in the previous step except that another parameter, α_4. AC is added to the previous cost function to minimize the weighted cost of the area, the wire lengths, and the attack correctness. AC value is determined by executing the proximity attack algorithm, discussed in Sect. 2.3, on the placement result at that point.

3.5.3   Discussion

Xie et al. [11] have evaluated the performance of their method using the ISCAS-85 and ITC-99 benchmarks. Setting $A_t = 0.1$ and varying S_t from 1.01 to 1.3, the cutset size decreased for all the benchmarks as S_t was increased. Secure min-cut was defined as the partitioning with S_t that makes HD larger than 40%. For normal placement $\alpha_1 = 0.2$, $\alpha_2 = 0.7$, $\alpha_3 = 0.1$, and $\alpha_4 = 0$ were used, whereas for secure placement these values were set to 0.2, 0.7, 0.05, and 0.05, respectively. With these settings, they have observed that normal partitioning and placement yield proximity attack correctness (AC) of 20.13% and HD of 11.98% on average. With normal partitioning followed by secure placement, these values changed to 0.22% and 13.24%, respectively. With secure partitioning followed by normal placement, AC of 9% and HD of 43.87% were observed. Finally, with secure partitioning and placement, the AC was 0.27% and the HD was 46.35%. These results show that these secure design methods lead to low attack correctness while ensuring near-optimal HD. Area and wire length overheads with the secure methods were, on average, 8.95% and 17.27%, respectively. However, the improved security comes at the expense of the cutset size which directly impacts the BEOL fabrication cost. For one benchmark, they have reported that, as S_t was reduced from 2 to 1.2, the cut size increased from 25 to 155, and the HD increased from 5.46% to 48.55%, while AC remained at 0% at the expense of area increase from 0.74% to 6.91% and wire length increase from 0.91% to 14.15%.

The low AC values combined with near-50% HD values indicate that the method offers a solid defense against the greedy proximity attack. However, how well this method can defend against enhanced forms of proximity attacks, such as those discussed in Sects. 2.4 and 2.5, is unclear. This method did not need swap pins since AC is directly considered as part of the cost function in the secure placement step and is computed by invoking the proximity attack for every contemplated SA move. Thus, secure placement has a significant impact on the quality of the final output at the expense of computation time.

3.6 Secure Multiway Min-Cut Partitioning

Whereas the previous section discussed secure bipartitioning, Chen et al. [17] proposed to identify the BEOL signals using a *multiway partitioning* algorithm that used security-aware weights attached to the nets while identifying the cutset.

3.6.1 Objective

Let $G = (V, E)$ be a hypergraph representation of a combinational logic circuit. Vertices v_i, $1 \leq i \leq |V|$ represent logic gates, and hyperedges $e_j \subseteq V$, $1 \leq j \leq |E|$ represent signal nets. Let $H \subseteq E$ be the subset of nets selected to be assigned to the hidden BEOL layers and fabricated at a trusted facility. Remaining circuit $(V, E - H)$ is assigned to the FEOL layers fabricated at a foundry assumed to be untrusted. The attacker knows $(V, E - H)$. Let H' be the set of nets that the attacker reconstructed in place of H. Hence, the attacker's final circuit is $G' = (V, (E-H) \cup H')$. The problem is to identify a suitable H so that it is computationally hard for the attacker to reconstruct the original circuit. Removing H from G induces a partition $\{V_1, V_2, \ldots, V_p\}$ on V such that $V = \cup_{i=1,p} V_i$ and $H = \{e \in E | \exists_{v_1,v_2 \in e} : v_1 \in V_i, v_2 \in V_j, i \neq j\}$. Conversely a partition of V into $\{V_1, V_2, \ldots, V_p\}$ for some p, $2 \leq p \leq |V|$ determines the "cutset" of nets H to be assigned to the BEOL layers. The problem, therefore, is to perform a security-driven multiway partitioning (for a suitable value of p) so as to identify and hide the cutset of nets in the BEOL layers to defeat the proximity attack.

3.6.2 Secure Partitioning

Security is considered by attaching suitably defined weights to the nets. Three factors contribute to the weight of a net: Signal Effect on Output, Output Cone Size, and Signal Priority.

1. *Effect of Signals on Primary Outputs:* A small ANHD value indicates a relatively high degree of matching with the correct circuit. Signals that are likely to cause multiple errors in the primary outputs thereby increasing the ANHD value are

preferred for embedding in the BEOL layers. To quantify the effect of signals on the primary outputs, a parameter named *Signal Effect on Output (SEO)* of a signal is defined. Let s be a signal net. SEO of a signal s in the context of an input vector x is defined as

$$\text{SEO}(s, x) = \text{HD}(F(x)|_{s=b}, F(x)|_{s \leftarrow b'}) \tag{3.10}$$

where $F(x)$ denotes the output vector when x is applied as input to the circuit, $b(\in \{0, 1\})$ denotes the value assumed by s when x is applied, and $F(x)|_{s \leftarrow b'}$ denotes the output vector when the value of s is reversed and forward propagated to the primary outputs while keeping the other values. Thus, SEO(s,x) denotes the sensitivity of the primary outputs to a change in s from b to b' in the "context" of x. Note that the value of SEO is dependent on x and the direction of change in s. To eliminate the bias of context x, SEO is averaged over a large number of input vectors. Given a series of input vectors $x_1, x_2, \ldots x_n$, SEO(s) is defined as

$$\text{SEO}(s) = \frac{\sum_{i=1}^{n} \text{SEO}(s, x_i)}{n} \tag{3.11}$$

Average SEO (ASEO) is the average SEO value of all the signals in the design. Finally, *Normalized SEO*, NSEO(s), of a signal s is defined as

$$\text{NSEO}(s) = \frac{\text{SEO}(s)}{\text{ASEO}} \tag{3.12}$$

2. *Primary Output Cone Size:* Consider two signals s_1 and s_2 that have the same *SEO* value over the same input vectors. It is possible that $SEO(s_1)$ value is due to its high influence on one pin or relatively few pins, while $SEO(s_2)$ value is due to its moderate influence on a large number of pins. To maximize ANHD, preference should be given to s_2 for inclusion in the BEOL layers. To introduce this weight, the *Primary Output Cone*, POC(s), of a signal s is defined as the set of primary output pins in the output cone of signal s. $|POC(s)|$ is the number of output pins influenced by s. Average Output Cone Size (AOCS) is the average value of $|POC(s)|$ over all the signals in the design. *Normalized Output Cone Size*, NOCS(s), of s is defined as

$$\text{NOCS}(s) = \frac{|\text{POC}(s)|}{\text{AOCS}} \tag{3.13}$$

3. *Priority Factor:* Consider a signal s that influences only one primary output pin p. Since, in this case, both NSEO(s) and NOCS(s) will be low, it is unlikely that s would be selected for inclusion in the BEOL layers if only these two metrics are considered. However, if s is the only signal that influences p, then s should be given a higher priority for inclusion in the BEOL layers in order to ensure that

---

Algorithm 9: Security weights of signal nets in a netlist [17]
___

Input: Signal Nets $S = \{s_1, s_2, \ldots, s_k\}$, Input Vectors $X = \{x_1, x_2, \ldots, x_n\}$, Primary
       Outputs $O = \{O_1, O_2, \ldots, O_w\}$

    Output: $SEO[1..k], POC[1..k], P[1..k]$

 1   **for** i **in** 1 **to** n **do**
 2   Simulate the circuit with input x_i and obtain the output vector y;
 3   Backup the internal signal values;
 4      $\forall j \in [1..k], \forall m \in [1..w], SEO_j[m] = 0$;
 5   **for** j **in** 1 **to** k **do**
 6   Propagate_Signal_Forward($s_j = \text{INV}(s_j)$);
 7   **for** m **in** 1 **to** w **do**
 8   **if** $(y[m] \neq Current\_Output\_Value(O_m))$ **then**
 9               $SEO_j[m] = SEO_j[m] + 1$;
10   **end**
11   **end**
12   Restore the internal signal values;
13   **end**
14 **end**
15 $\forall j \in [1..k], POC[j] = 0$;
16 $\forall j \in [1..k], SEO[j] = 0$;
17 **for** j **in** 1 **to** k **do**
18   **for** m **in** 1 **to** w **do**
19   **if** $(SEO_j[m] \neq 0)$ **then**
20            $POC[j] = POC[j] \cup \{O_m\}$;
21            $SEO[j] = SEO[j] + SEO_j[m]$;
22            $Q[y] = Q[y] + 1$;
23   **end**
24   **end**
25 **end**
26 $\forall j \in [1..k], P[j] = 0$;
27 **for** j **in** 1 **to** k **do**
28   **foreach** $i \in POC[j]$ **do**
29         $P[j] = \min(P[j], Q[i])$;
30   **end**
31 **end**
32 **return** SEO, POC, P;

---

p contributes to the ANHD of an incorrectly reconstructed circuit. To generalize this situation, we introduce a *Priority Factor*, P, of signal s, defined as follows:

$$P(s) = \underset{p \in POC(s)}{MIN} |\{i : signal \mid p \in POC(i)\}| \tag{3.14}$$

The lower the priority factor, the higher is the preference to assign the signal to BEOL layers.

The procedure to calculate these factors is shown as Algorithm 9. After taking all parameters into consideration, a set of highly influential signals is selected for BEOL fabrication.

A hypergraph partitioning tool, PaToH [18], is adapted for secure partitioning. PaToH is a multilevel hypergraph partitioning tool based on the Fiduccia–Mattheyses (FM) partitioning algorithm [12] discussed in Sect. 3.5. It generates near-optimal multiway partition results for hypergraphs with weighted vertices and edges. The three parameters defined in the previous subsection are used to weigh the nets. Then, the weighted min-cut partitioning of PaToH is used to generate balanced p-way partitions for various p values.

Since heavily weighted nets are unlikely to be cut, to each signal net, s is assigned a weight that is directly proportional to P(s) and inversely proportional to NSEO(s) and NOCS(s) as shown below:

$$W(s) = \begin{cases} \infty & \text{if } |POC(s)| = 0 \\ \alpha * \frac{P(s)}{NSEO(s)*NOCS(s)} & \text{otherwise} \end{cases} \qquad (3.15)$$

α is the proportionality constant which is tuned experimentally. For the signals that do not effect the primary outputs, the weights are set to the maximum value allowed by PaToH. Proximity attack aims to recover the BEOL nets based on the assumption that nearby FEOL circuit pins are more likely to be connected. The aim is to defeat the proximity attack by increasing the number of partitions that together form the FEOL circuit and assigning the security-aware cut nets to BEOL layers. Furthermore, as the number of partitions increases, proximity attack often fails to identify a cycle-free connection with a target pin since viable candidate pins have already been incorrectly assigned to other nets.

3.6.3   Global Placement for Minimum Wire Length

Pin placement is critical to defeat location based attack algorithms. If the same weights used in partitioning are used to drive placement, then pins connected by the BEOL nets tend to be placed close to each other thereby allowing the proximity attack to succeed. In order to thwart the attack, as discussed in Sect. 3.4, pin positions are usually swapped or perturbed thereby increasing wire length and possibly adversely effecting the performance. However, in this approach, such perturbation is not necessary due to the use of different weight functions during partitioning and floor-planning/placement. Without loss of generality, the circuit is assumed to be placed on a square die and partitions are assigned to a square grid on the die. The number of partitions p is assumed to be a perfect square of some integer ≥ 1, that is, $p = 2, 4, 9, 16 \ldots$. The hypergraph partitioning tool evenly divides the circuit into p balanced partitions.

For a multiway partitioned circuit, the location of each partition will affect the wire length dramatically. Hence, Chen et al. first perform a floor-planning of the p partitions to minimize the total global wire length. After partitioning, a simulated annealing (SA) based min-weight floor-planning algorithm allocates space for each partition. Finally, a global placement of the whole circuit is performed using a constrained force-directed placement algorithm in which the cells in each partition are constrained to stay within the area assigned to that partition. In the force-directed placement algorithm [19], the ideal target location for any cell can be anywhere in the designated placement area. However, in this case, if the target location for any cell is computed to be outside the region designated for its partition, then it is changed to the nearest location within the partition's region.

3.6.4 Discussion

Chen et al. [17] used benchmarks from the ISCAS-85 and ITC-99 suites to evaluate the secure multiway partitioning and placement method. They have reported modest reduction in bipartitioning cutset sizes to achieve the same level of security as the method discussed in Sect. 3.5. For all but three large benchmarks, >40% HD was achieved with $p = 2$. For the three relative large benchmarks, 16 or more partitions were required so that a large number of appropriate BEOL signals could be identified. AC values remained at <5%. Without the security-driven net weights, the HD dropped below 10%, whereas using the weights resulted in a HD of >40%, which shows their effectiveness. Experiments showed that for large benchmarks, the HD increases and AC decreases as the number of partitions increases. This is expected due to the increase in the cutset size. However, as the number of partitions increased exponentially, the cutset size increased linearly, that is, the BEOL fabrication cost increased linearly. The total HPWL overhead remained below 10% even for a very large number of partitions and, in some cases, actually decreased by up to 8.24%.

Similar to the secure bipartitioning method, Chen et al.'s method aims to thwart the greedy proximity attack but not the extended proximity and network flow attacks.

The secure partitioning and placement methods, discussed in Sects. 3.5 and 3.6, show that it is possible to defeat the simple proximity attack by a careful selection of the cut nets while using traditional partitioning algorithms with only a modest increase in the cost metrics.

3.7 Placement Perturbation

The defense methods discussed so far defend against the greedy proximity attack but not necessarily against the network flow attack discussed in Sect. 2.5. Wang et al. [20] proposed a method to make small changes to the placement to defend against attacks that exploit physical design information, including the network flow attack.

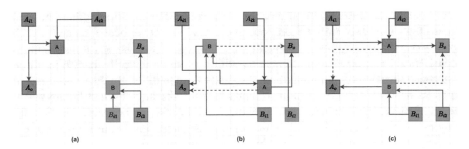

Fig. 3.2 Placement perturbation (based on [20]). (**a**) Example placement, (**b**) Placement after swapping A and B, (**c**) Placement after perturbing A and B

Their method takes the wire length into account while making small changes to the locations of selected gates in a placed and routed design.

3.7.1 Objective

Consider the example shown in Fig. 3.2a where the total wire length is 19 units. Gates A and B are swapped in Fig. 3.2b, and the dotted wires are assigned to BEOL to defeat the proximity based attacks. Red wires are the incorrect connections that a proximity attack or network flow attack would recover. However, the wire length increased by 126% to 43 units. Figure 3.2c shows that the attack can be defeated by making small perturbations to the positions of A and B instead of swapping. In this case, the wire length increased only by 42% to 27 units. This shows that small perturbations to selected gates could be sufficient to mislead the attackers.

The method of perturbation has two steps, *gate selection* and *placement pertur-bation*. In the first step, a set of tree subcircuits is extracted from the design and, in the second step, their locations are perturbed.

3.7.2 Gate Selection

Two methods to extract suitable trees of gates are proposed as follows:

1. *BEOL-Driven Gate Selection (BGS):* This method, shown as Algorithm 10, selects gates that are connected to the BEOL wires such that moving those gates would confuse the attackers trying to recover the BEOL wires. Given a circuit in which the BEOL nets are already identified, the algorithm returns several trees of gates such that each BEOL wire is incident on one of the selected gates.
   The algorithm begins by a reverse topological sort of the circuit and assigning a level number $R(g)$ to every gate g in the circuit (line 2). In each iteration, a gate

---

Algorithm 10: BEOL-driven gate selection (BGS) [20]

Input: Circuit C, BEOL Nets W

Output: Set of Tree Subcircuits $\mathcal{T}$

1 Set of trees $\mathcal{T} = \emptyset$;
2 Determine the reverse topological level number $R(g)$ for each gate g in C;
3 G = Set of all gates in C which have a BEOL wire as input;
4 **while** $G \neq \emptyset$ **do**
5   Select a gate $r \in G$ such that r has the least possible level number;
6   Tree $T = \{r\}$;
7     $G = G - \{r\}$;
8   **while** $\exists g \in T, \exists (d, g) \in W, d \notin T$ **do**
9   Select a gate $g \in T$ with the least possible level number such that $\exists (d, g) \in W, d \notin T, R(d) = R(g) + 1$;
10         $T = T \cup d$;
11         $G = G - \{d\}$;
12   **end**
13     $\mathcal{T} = \mathcal{T} \cup \{T\}$;
14 **end**
15 **return** $\mathcal{T}$;

---

r driven by a BEOL net is selected from the lowest level where such a gate is available (line 5). A new tree is constructed with r as its root (line 6). The tree is expanded by iteratively traversing the circuit from r in the reverse topological order and selecting all gates connected by BEOL wires until either primary inputs or gates driven only by non-BEOL wires are reached (lines 8–12). In addition, for multi-fanout gates, only the first fanout gate encountered is included to ensure that the selected subcircuit has a tree topology (not shown in the algorithm).

2. *Logic-Aware Gate Selection (LGS):* This method is based on two observations: (1) Following placement perturbation, incremental routing will be performed. Due to this, some layer assignment of wires could change, and it is possible that a gate that was incident to only FEOL wires would become incident to wires now assigned to the BEOL layers. (2) If two gates that produce similar outputs are interchanged due to wrong connections by an attacker, it is unlikely to have much impact on the circuit outputs. Logical dissimilarity of the gates impacts security. These observations motivate logic-aware gate selection. This method also produces trees of gates as before but with a couple of important differences. In addition to gates incident to BEOL nets, gates incident to the top FEOL layers are included in the traversal. However, in order to contain the wire length overhead due to the larger trees, only some gates in the tree are allowed to move, while others remain fixed. *Logical dissimilarity*, defined as the probability that two gates produce opposite values, is used to determine the mobility of the gates. These probabilities can be determined by simulations. Two gates separated by a given distance are defined as *neighbor gates with significant logical difference* (NGSLD) provided their logical dissimilarity is above a given threshold Φ.

A relatively small Φ is used for gates incident to BEOL wires since these gates
are more likely to be incident to BEOL wires after rerouting, and a larger Φ
is used for gates that are incident only to FEOL layers since these gates are
less likely to be incident to BEOL wires after rerouting. All gates crossing these
thresholds are considered and sorted by non-increasing order of their number
of NGSLD neighbors. Then the top $\rho\%$ of them are selected as movable gates,
while the remaining are fixed gates. ρ parameter impacts the tradeoff between
security and wire length. Larger ρ results in more gates being perturbed which
generally improves security at the expense of wire length.

3.7.3 Placement Perturbation

Two methods to perturb the placements of the selected gates in each extracted tree
are suggested:

1. *Physical-Driven Placement Perturbation (PPP):* Each gate in the tree is perturbed starting with the leaf gates and proceeding in the topological order toward
   the root. For each leaf gate, a set of candidate solutions is obtained by varying its
   location. Leaf candidate solutions are propagated to the parent nodes and are
   merged at the parent nodes. Each candidate solution is evaluated by its wire
   length increase and a *perturbation metric* that is defined as follows:
   Consider a net with a source gate S and sink gates S_i. When these gates are
   perturbed, not only the distances between them change but also their spatial
   order (relative positioning on the placement surface) may change. Perturbation
   between a source S and a sink S_i in x-direction is defined as the product of a
   perturbation gain factor and change of distance between them in the x-direction.
   The gain factor is one (zero), if the spatial order remained the same in the x-
   direction and the x-distance increased (decreased). The gain factor is 2 if the
   spatial order flipped in the x-direction. Perturbation in the y-direction is similarly
   defined. Perturbation of S, π_S, is the sum of the perturbations of all (S, S_i) pairs
   in both x- and y-directions. A solution with large perturbation is more difficult
   to attack. Each candidate solution is characterized by its wire length w and
   perturbation π.
   Consider the example placement shown in Fig. 3.3a and its perturbation in
   Fig. 3.3b. For source A and sink C, along the y-axis, the perturbation magnitude
   is $|-10-10| = 20$. In addition, the spatial order has flipped; C is now to
   the north of A. Hence, a gain factor of 2 is applied, resulting in perturbation
   $\pi_{A,C}^y = 2 \times 20 = 40$. For source A and sink C, along the x-axis, the distance
   decreased and there is no spatial order change. Hence, the gain factor is 0
   since this does not change the proximity between A and C in the x-direction.
   Hence, $\pi_{A,C}^x = 0.|20-15| = 0$. For source A and sink B, the x-distance
   increased without changing the spatial order. The gain factor in this case is

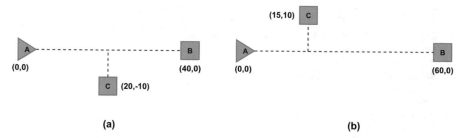

Fig. 3.3 Placement perturbation metric (based on [20]). (**a**) Original placement, (**b**) Perturbed placement

1. Hence, $\pi_{A,B}^{x} = 1.|40 - 60| = 20$. The overall gain factor for this net is $\pi_A = \pi_{A,B}^{x} + \pi_{A,B}^{y} + \pi_{A,C}^{x} + \pi_{A,C}^{y} = 20 + 0 + 0 + 40 = 60$.

   At each tree node, some inferior solutions can be pruned without further propagation. Given two solutions s_1 and s_2, s_1 is inferior to s_2 provided $w_1 \geq w_2$ and $\pi_1 \leq \pi_2$. Once all the solutions at the root node are determined, the solution that does not exceed a specified wire length overhead factor α and has maximum perturbation is selected.

   The physical-driven placement perturbation method is shown as Algorithm 11. Without loss of generality, the algorithm assumes a binary tree. While moving the gates, each gate can be moved to a vacant location in the vicinity or gates within a tree can be moved to each other's locations.

2. *Logic-Driven Placement Perturbation (LPP):* In this method, the perturbation metric is replaced by the *weighted logical difference* (WLD) metric which takes into account the logical difference among the gates being perturbed. Given a gate g being placed at a temporary location l, WLD is defined as

$$\text{WLD}(g, l) = \lambda(g, l) \sum_{h \in G_l} \frac{\Delta(g, h)}{dist(g, h)} \tag{3.16}$$

where $\lambda(g, l)$ is the ratio of wire length related to g at location l to the original wire length, G_l is the set of critical gates around l, $\Delta(g, h)$ is the logical difference between g and h, and $dist(g, h)$ is the rectilinear distance between the location l and the location of h.

Consider the example in Fig. 3.4. Let A (at location l) and B be the source and sink nodes of a BEOL net. Assume that unrelated gates C and D also have their output nets in the BEOL. B is the closest sink node to A. Hence, an attacker can easily recover the BEOL net from A to B. If A were to be moved to location l', in the vicinity of C and D, then the attacker cannot immediately determine which source among A, C, and D is the correct driver for B. WLD of A at l' is

$$\text{WLD}(A, l') = \frac{dist(B, l')}{dist(B, l)} \left(\frac{\Delta(A, C)}{dist(C, l')} + \frac{\Delta(A, D)}{dist(D, l')} \right)$$

Algorithm 11: Physical-driven placement perturbation (PPP) [20]

Input: Tree of Gates T, Wire Length Increase Factor α
Output: Locations of gates in T
1 w_{ini} = wire length of T;
2 **for** *every fanin i of every leaf node of T* **do**
3 | set a solution for i with $(w = 0; \pi = 0)$;
4 **end**
5 **for** *each gate $g \in T$ in topological order* **do**
6 | Fan-in solution set $S = \emptyset$;
7 | (g_1, g_2) = fanin of g;
8 | **for** *each perturbation solution s_1 of g_1* **do**
9 | **for** *each perturbation solution s_2 of g_2* **do**
10 | | $S = S \cup \{(w_1 + w_2, \pi_1 + \pi_2)\}$;
11 | **end**
12 | **end**
13 | Solution set of g, $S_g = \emptyset$;
14 | **for** *each candidate location (x_i, y_i) of g* **do**
15 | Temporarily place g at (x_i, y_i);
16 | Solution set for the candidate location, $S_{g,i} = \emptyset$;
17 | **for** *each solution $s \in S$* **do**
18 | | Obtain (w_i, π_i) based on (x_i, y_i) and s;
19 | | $S_{g,i} = S_{g,i} \cup \{(w_i, \pi_i)\}$;
20 | **end**
21 | Prune $S_{g,i}$;
22 | $S_g = S_g \cup S_{g,i}$;
23 | **end**
24 **end**
25 S_{root} = solutions for the root gate of T;
26 Find a solution $s \in S_{root}$ that has maximum π and $w \leq (1 + \alpha).w_{ini}$;
27 **return** location of each gate in T according to s;

Fig. 3.4 Logic-driven
perturbation (based on [20])

Solutions with large values of WLD are retained for further exploration. This prefers solutions with large logical difference from its neighbors thereby increasing security. The algorithm for logic-driven placement perturbation is similar to Algorithm 11 except for the use of the WLD metric in place of the perturbation metric.

3.7.4 Discussion

Wang et al. [20] evaluated the placement perturbation methods to defend against the network flow attack, discussed in Sect. 2.5, using the ISCAS-85 and ITC-99 benchmarks. Three combinations of methods are evaluated: BGS + PPP, LGS + PPP, and LGS + LPP. When wire length overhead is limited to a modest 10%, all three defense methods resulted in lower correct connection recovery rates using the network flow attack compared with no defense method being used. However, the remarkable gain was in the output error rate (OER). The three combinations showed progressive improvement in OER, with the LGS + PPP method yielding more than 80% in some cases.

Average wire length overhead was 5.1%, 5.7%, and 3.2% for BGS + PPP, LGS + PPP, and LGS + LPP, respectively. Increase in critical path delay was less than 0.5% in most cases, usually correlating with the wire length. Power increase showed as similar trend. However, critical path delay and power are not directly controlled by the perturbation methods.

Security vs overhead tradeoff study showed that both ρ and α can be effectively used to control this tradeoff. Furthermore, LGS + PPP and LGS + LPP methods were shown to be superior to BGS+PPP as the logic-based methods can decrease the correct connection rate (CCR) and increase the error rate with lower wire length overhead.

Finally, experiments showed that the network flow attack can be ineffective when splitting is done at lower metal layers (with most nets in BEOL) and that the perturbation defenses are ineffective when splitting is done at upper layers (with very few nets in BEOL). The defense methods work best when designers allow a few lower levels of metal to be in FEOL and delegate the rest to BEOL.

Wang et al.'s methods showed the importance of considering logic value-based decisions when constructing a defense. We will see that the logical differences among pins are used in several forthcoming defense methods.

3.8 Routing Perturbation

The placement perturbation method thwarts the network flow attack by directly controlling placement decisions which indirectly influences routing. Recall from the discussion in Sect. 2.4 that routing has more influence than placement on deterring proximity attacks. Wang et al. [21] proposed a *routing perturbation* method to defend against the proximity and network flow attacks. This method allows more direct control of the wire length overhead than the placement perturbation method.

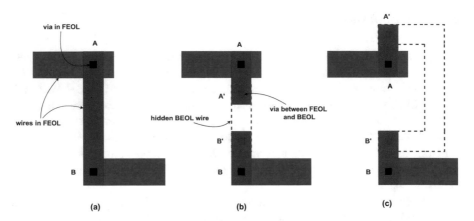

Fig. 3.5 Layer elevation and routing detour (based on [21]). (**a**) Example net AB, (**b**) Split of net AB, (**c**) Routing detour

3.8.1 Objective

The method proposed by Wang et al. is based on four basic techniques for routing perturbation:

1. *Layer Elevation:* In general, moving more nets from FEOL to BEOL layers enhances security. Figure 3.5a shows a net AB in a layout in FEOL layers. Figure 3.5b shows the net split into three fragments, AA', BB', and A'B'. The first two fragments remained in FEOL, while the third one is elevated to BEOL by introducing FEOL-BEOL vias on both ends. On the FEOL layers, the vias appear as "dangling" pins to an attacker who needs to reconstruct the missing connectivity between them. However, unless further changes to the layout are made, the attackers can use proximity and other physical information to recover the BEOL nets.

2. *Routing Detours:* Proximity attack is based on the assumption that a wire connecting two pins is normally routed through the shortest path. Wires can be detoured to confuse the proximity attack at the expense of wire length and, possibly, performance and power. Figure 3.5c shows a detour introduced in the AA' fragment of the net. Since dangling pins A' and B' are no longer close to each other, proximity based attacks are unlikely to succeed.

3. *Decoy Usage:* Even when wires are rerouted to take longer paths to increase the distance between the two dangling pins, if the two pins have no other unconnected pins nearby, then the proximity attack can still connect them. Hence, in addition to detouring, the dangling pin should be placed in the vicinity of other unrelated dangling pins. The presence of several unconnected pins in the neighborhood will confuse proximity based attacks. In Fig. 3.6, dangling pin A', which was originally to the south of A, is moved to the east of A where it is in

Fig. 3.6 Decoy wires (based on [21])

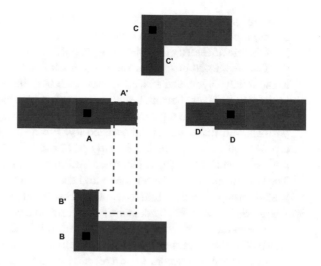

the proximity of unrelated dangling pins C' and D'. Proximity based attacks will be tempted to connect A' to either C' or D' instead of B'.

4. *Testability:* Routing perturbation, while minimizing the wire length overhead, should maximize security. The Hamming distance (HD) metric is used to quantify the level of security attained. To increase HD, fault observability defined by the Sandia Controllability/Observability Analysis Program (SCOAP) [22] is used as a metric. Nets with high observability are preferred for perturbation. Logical difference discussed in Sect. 3.7 is used as a metric to select decoys for a pin. Logical differences are determined using logic simulations.

The objective of the routing perturbation method is to use these four basic techniques to improve security while minimizing wire length.

3.8.2 Algorithm

Routing perturbation algorithm uses these ideas to select some nets, rip up a portion of their wire connections, make changes to the remaining fragments, and reroute the fragments that were ripped up. The algorithm has the five phases discussed below:

1. *Ripping up BEOL wire segments:* In the first phase, all BEOL nets (nets that have some wire fragments assigned to the BEOL layers) are sorted in non-increasing order of their SCOAP observability values. From this, the top $\rho_B\%$ of nets are selected for partial rip up and reroute in the next phases. ρ_B controls wire length overhead and security tradeoff. Each BEOL net has a portion assigned to the BEOL layers, while the rest remains in the FEOL layers. The BEOL part is ripped up leaving two dangling terminals in the top FEOL layer. These terminals are called *pseudo pins*.

2. *Ripping up FEOL nets:* All FEOL nets (nets that are routed entirely in the FEOL layers) are sorted in non-increasing order of their SCOAP observability values. Nets that do not satisfy the following properties are discarded from consideration: (1) The net should have a wire segment such that there exists some empty space in the BEOL layers and space for vias such that the wire segment can be routed in the BEOL layer. This makes layer elevation feasible. (2) The net should have wire segments in the top-most FEOL layer. This makes the perturbation easy to perform and likely performance friendly. From the remaining FEOL nets, the top $\rho_F\%$ of nets are selected for rip up. In each net selected, a wire segment in the top-most FEOL layer is identified and removed resulting in two pseudo pins. The two terminals will be relocated and the wire fragment between them will be routed through the BEOL layers in the subsequent steps.

3. *Driver side detour:* Each net identified for rip up has two pseudo pins, one toward the driver side and the other toward the sink side of the net. In this step, the driver side FEOL wire fragment connecting the driven to the pseudo pin will be rerouted by moving the location of the driver side pseudo pin from its current side (say, east) to the opposite side (west) of the driver such that it is as close as possible to the driver and has space to be connected to the BEOL layers. Detouring confuses the attacker and proximity to the driver reduces wire length overhead.

4. *Sink side decoy:* In this phase, sink side pseudo pins introduced in the rip up steps will be relocated to be in the vicinity of high-quality decoys. For each net, this is done in two steps. First, change the location of the sink side pseudo pin p from its current side (say, east) to the opposite side (west) of the sink node. Then, on that side, it will be positioned close to multiple decoy pins. Several alternative locations are evaluated as follows: For each location, the nearest driver side pseudo pin (potential decoy) of an unrelated net is identified. Let the distance between this pin and p is d. Any other decoy pins in the ring centered on p with distance $d + \Delta d$ are identified. Δd is selected such that the sink node or the parent node of p is in the ring. A driver side pseudo pin in this ring is a decoy if its connection with p does not force any wire detours, that is, a proximity attack is highly likely to connect them. The average logical difference (ALD) between these decoys and the original driver of p is determined using simulation data. From the candidate locations for p, the location that offers the maximum ALD is selected as the new location of p.

   Consider Fig. 3.7. AA' is the net being primed for protection. The sink side pseudo pin A' is being moved from the south of A to the north of A. One candidate location for A' is shown. D' is the nearest driver side pseudo pin to A'. Let d be the distance between A' and D'. A circle with radius d centered at A' is drawn. Another circle with radius $d + \Delta d$ is also drawn such that the sink pin of A' is in the ring formed by the two circles. Now, unrelated source side pseudo pins B' and D' happened to be in this ring. Among these, B' is not in the direction of the dangling wire AA' and is excluded from consideration since network flow type attacks omit such pins from the candidate list of possible matches for A'. The remaining pseudo pins, C' and D', constitute legitimate decoys for A'.

Fig. 3.7 Sink side decoy
(based on [21])

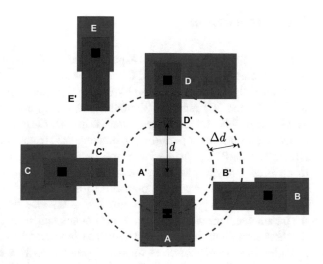

Algorithm 12: Routing perturbation [21]

Input: Layout L, Netlist N, Control Parameters ρ_B, ρ_F
Output: Modified Layout L
1 Obtain SCOAP observability values of all nets in N;
2 Sort BEOL nets in non-increasing order of their observability values;
3 N_R = select the top $\rho_B\%$ of the BEOL nets;
4 **for** *each net* $n \in N_R$ **do**
5 | Rip up the BEOL wire segments from n;
6 **end**
7 Sort FEOL nets in non-increasing order of their observability values;
8 Discard non-qualifying FEOL nets from the sorted list;
9 N_F = select the top $\rho_B\%$ of the FEOL nets;
10 **for** *each net* $n \in N_F$ **do**
11 | Rip up a wire segment in the top-most FEOL layer from n;
12 **end**
13 $N_R = N_R \cup N_F$;
14 **for** *each net* $n \in N_R$ **do**
15 | Move the driver side pseudo pin to a suitable location on the other side of the driver;
16 | Move the sink side pseudo pin to a location on the other side of the sink such that its
   | average logical difference with the decoy pins in the neighborhood is maximized;
17 **end**
18 L = Reroute all the ripped up wire fragments between the pseudo pins in the BEOL layers;
19 **return** Modified Layout L;

---

5. *BEOL wire rerouting:* The ripped up wire fragments connecting the respective
   pseudo pins of all the nets are now rerouted through BEOL layers using any
   available router.

The routing perturbation algorithm is summarized as Algorithm 12.

3.8.3   Discussion

Wang et al. [21] have conducted experimental studies using the ISCAS-85 and ITC-99 benchmarks. They have used the network flow attack [23] discussed in Sect. 2.5 to evaluate the security afforded by the routing perturbation method based on the ICR, OER, and HD metrics. Without any defense, the attack achieved ICR as low as 11% (i.e. 89% attack guesses were correct). OER and HD were 73% and 8% on average. Placement perturbation defense [23] (Sect. 3.7) yielded ICR, OER, and HD of 12.5%, 87%, and 13.5%, respectively. Routing perturbation improved ICR to 36% on average and improved OER to 100% and HD to 27%. Wire length overhead was less than 3% compared to about 6% with placement perturbation. Increase in critical path delay was nominal at 0.23%. Experiments also showed that the parameter ρ provided smooth control over security vs overhead tradeoff. Among the three techniques incorporated in the routing perturbation method, layer elevation provided most contribution to improved security as it is responsible for increasing the number of hidden BEOL nets. Detour and decoy techniques work best when used in conjunction with layer elevation.

Similar to the placement perturbation method, the routing perturbation method considered logical differences among terminals while making changes to the layout. This plays a significant role in obtaining good HD and OER results while containing the wire length overhead.

3.9   Concerted Wire Lifting

The defense methods discussed so far underline the importance of selective net lifting into BEOL layers. Patnaik et al. [24] proposed the method of concerted *wire lifting* to move selected nets to the BEOL layers and rerouting them while mitigating proximity hints.

3.9.1   Objective

In this method, rerouting of the lifted nets is facilitated by the use of custom cells, called *elevating cells* (ECs), which establish pins in the selected BEOL layers and, optionally, provide dummy pins to confuse the attacker. Figure 3.8 illustrates the idea of an elevating cell. Figure 3.8a shows a net routed through five metal layers. M5 is considered the only BEOL layer. Both top view and side view are shown. The wire fragment in M5 is hidden from the attacker. In this case, since the routing is done by an automated router that tries to accomplish as much routing as possible in

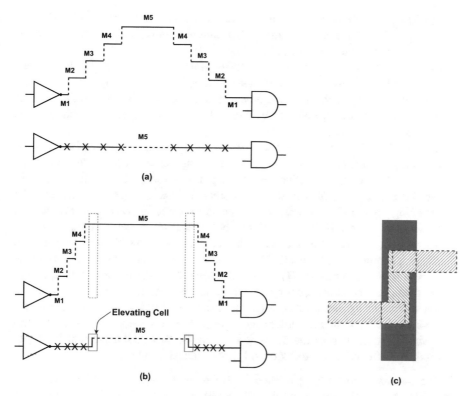

Fig. 3.8 Elevating cell (EC) (based on [24]). (**a**) Example net, (**b**) Elevating cell, (**c**) Layout view of EC

lower layers, only a short wire fragment is left in M5 and the short distance between the two dangling pins in M4 is determined by the router. Figure 3.8b shows an EC. The cell contains M5 wiring. The net is split at any desired points on the source and sink sides and ECs are on both sides. The distance between the ECs can be controlled as precisely as desired. Increasing the distance between ECs increases the M5 wire length and makes proximity based attacks difficult. Figure 3.8c shows a conceptual layout view of the EC.

The objective of the concerted wire lifting method is to determine appropriate nets to be lifted and determine the locations for EC insertion so as to minimize wire length while maximizing security.

3.9.2   Algorithm

Concerted wire lifting uses three basic techniques:

1. *Lifting high fanout nets:* Nets with two or more sinks are selected, and all branch wires are lifted to result in as many dangling pins as possible. ECs are inserted for the driver and for each of the sinks.
2. *Controlling the distance between dangling pins:* ECs introduced in the previous step are placed close to the source and sink nodes, respectively. This increases the distance between the dangling pins introduced by wire lifting and discourages proximity attacks. Placement of ECs can also be randomized to defeat machine learning based attacks.
3. *Obfuscation of short nets:* For short nets, enlarging the distances between the dangling pins is hard without introducing wiring detours. For lifting short nets, a special EC containing a dummy driver pin is introduced. This EC, shown in Fig. 3.9, contains two pins close to each other: a pin connected to the driver of the net being lifted and a *dummy* pin connected to the unrelated driver. Figure 3.9a shows the top and side views of the obfuscated short net and the unrelated driver, both connected to an EC at the source side of the drivers. Figure 3.9b shows a layout view of the EC. The unrelated driver is selected randomly while satisfying two conditions: (1) If the attacker selects the dummy driver, it will not result in a combinational cycle, which means that the proximity attack will not automatically discard it, and (2) Both the true driver and the dummy driver have similar drive strengths after accounting for routing detours. This will defend against attacks, such as the network flow attack, that take drive matching into account. Only one EC is introduced for the lifted short net.

The concerted wire lifting method is shown as Algorithm 13. While selecting nets (line 2), the algorithm initially selects high fanout nets and then long nets. Nets with larger fanout are given higher priority. Similarly, longer nets are given higher priority. Multiple ECs are inserted for these nets as discussed in the first two strategies. After this, short nets are selected until the PPA budget is exhausted. Shorter nets receive higher priority since those are vulnerable to the proximity attack and should be obfuscated. One special EC is inserted for each lifted short net.

Algorithm 13: Concerted wire lifting [24]

Input: Layout L, Netlist N, PPA Budget B
Output: Modified Layout L
1 **while** *PPA budget B allows additional lifting* **do**
2   Select a net n to lift;
3   Lift n and insert ECs in L;
4   Perform ECO optimization and legalization;
5   Reroute L, remove ECs;
6   Extract RC information and compute PPA data;
7 **end**
8 **return** Modified Layout L;

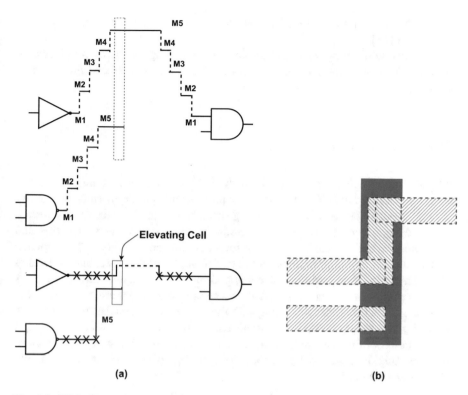

Fig. 3.9 EC for short wires (based on [24]). (**a**) Obfuscated short net, (**b**) Layout sketch

3.9.3 Discussion

Patnaik et al. [24] reported extensive experimental studies using the ISCAS-85 and the IBM superblue benchmark suites implemented using up to 10 metal layers with M6 as the split layer. They have used PNR (Percentage of Netlist Recovery) as the figure of merit. The number of dangling pin pairs introduced by concerted net lifting was shown to be far higher than both the original layouts and those induced by naive wire lifting. They have used the network flow attack, discussed in Sect. 2.5, as the reference attack.

When routing perturbation (Sect. 3.8) was used as the defense, network flow attack consistently resulted in 24% higher PNR on average compared with the use of concerted net lifting method as a defense to lift the same number of nets. Experiments showed that without any defense, attackers can recover, on average, 96% of nets. This drops to 95%, 88.5%, and 31% with the use of placement perturbation, routing perturbation, and concerted net lifting methods, respectively. Concerted net lifting also resulted in average in area, power, and delay of 9.2%, 10.7%, and 15%, respectively.

Final layouts for some of the benchmarks and EC implementations were made available in [25].

Concerted wire lifting achieved significant security improvement over placement and routing perturbation methods at the expense of delay overhead for the large benchmarks.

3.10   Netlist Clustering

In Sects. 3.5 and 3.6, we have discussed the use of circuit partitioning to identify the nets to be lifted to BEOL. Clustering is an alternative to partitioning to identify subcircuits such that the wires connecting these subcircuits are marked for lifting. In [26], Sengupta et al. suggested methods for clustering gates in a netlist and performing layout synthesis separately for each cluster. The goal is to separate connected gates by a larger distance so as to defeat proximity based attacks. During layout generation, the partitioned are mapped into layout regions called *fences*. A cluster is placed in a fence such that all the cells in the cluster are confined to the fence. This amounts to flour-planning with the clusters considered as modules. Two alternative clustering methods are proposed. Both methods use a *directed graph* representation of the given netlist, illustrated in Fig. 3.10. IO ports as well as gates are represented by nodes in the graph, and edges represent the connectivity among the gates and IO ports.

3.10.1   Gate Type Based Clustering

In this method, gates of the same type are grouped together. This is based on the observation that in designs, connectivity among the gates of the same type is relatively low. Gates performing the same function (e.g. NAND) are defined to be of the same type. Alternatively, type can be defined based on both function and fanin (e.g. two-input NAND) match. The second definition increases the number

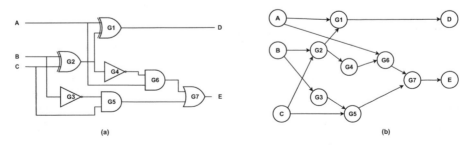

Fig. 3.10 Graph representation. (**a**) A netlist, (**b**) Corresponding graph (based on [26])

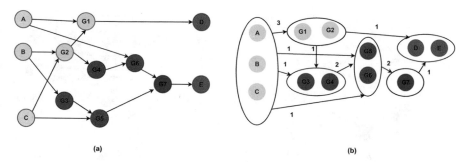

Fig. 3.11 Gate type based clustering (based on [26]). (**a**) Clusters shown by different colors, (**b**) Connectivity among the clusters

---

Algorithm 14: Gate type based netlist clustering [26]

Input: Netlist N
Output: Netlist Partitions $\mathcal{N}$
1 G = set of gates in N;
2 **for** *each gate type t except BUF or INV* **do**
3 | $N_t = \emptyset$;
4 **end**
5 **for** *each gate $g \in G$ with type t* **do**
6 | **if** $t = BUF$ *or* $t = INV$ **then**
7 | | Select a partition N_i uniformly randomly;
8 | | $N_i = N_i \cup \{g\}$;
9 | **else**
10 | | $N_t = N_t \cup \{g\}$;
11 | **end**
12 **end**
13 $\mathcal{N} = \cup_t \{N_t\}$;
14 **return** $\mathcal{N}$;

---

of partitions and is likely to increase security. Figure 3.11a shows an example clustering. Figure 3.11b shows the clusters formed and the system-level connectivity among the clusters.

The method, shown as Algorithm 14, groups gates of the same type together (line 10) except that the BUF and INV gates are placed in any of the other groups selected randomly (lines 6–8). Resulting cluster sizes may or may not be balanced depending on the distribution of gate types in the netlist. Usually, netlists contain a large number of BUF and INV cells that would form a dominant cluster. To avoid this, they are placed in other clusters randomly.

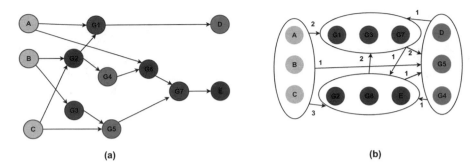

Fig. 3.12 Graph color based clustering (based on [26]). (**a**) Clusters shown in different colors, (**b**) Connectivity among the clusters

3.10.2 Graph Coloring Based Clustering

In this method, a graph coloring [27] algorithm is used for clustering. Each gate is assigned a color. Gates of the same color constitute a cluster. Initially, a gate is selected at random and assigned a color. Then, iteratively, the neighbors of an already colored gate g are assigned colors such that (1) each neighbor has a different color than that of g, (2) the colors of the neighbors are also different from those of one another, and (3) for each coloring assignment, the color that is used for the least number of gates so far is used. These choices ensure that connected cells are placed in different clusters and the cluster sizes are balanced. Figure 3.12a shows an example coloring, and Fig. 3.12b shows the system-level connectivity among the clusters formed.

Graph coloring based clustering method is shown as Algorithm 15. IO ports as well as gates are modeled as graph vertices for simplicity. Without loss of generality, the graph is assumed to be a connected graph. In the algorithm, N_c denotes the set of gates of color c and $C(g)$ denotes the color of gate g. In each coloring step (line 8), a suitable color, as defined above, is used. If no such color exists, then a new color, i.e. a new cluster, is generated.

3.10.3 Discussion

Sengupta et al. [26] have conducted experiments with the ISCAS-85 and MCNC benchmarks utilizing up to 10 metal layers. They have used the mutual information (MI) metric as the figure of merit and used the network flow attack, discussed in Sect. 2.5, to evaluate attack resiliency.

When splitting is done at M1, *random placement* of gates reduces MI by about 97% on average, while the clustering methods reduce it by about 88%. However, random placement results in a significant increase in wire length, up to 600%, while

---

Algorithm 15: Gate coloring based netlist clustering [26]

Input: Netlist N
Output: Netlist Partitions $\mathcal{N}$
1 G = set of gates in N;
2 Select a random gate $g \in G$;
3 Generate a new color c;
4 $N_c = \{g\}$;
5 $C(g) = c$;
6 **while** *there is a colored gate in g which has some uncolored neighbors* **do**
7   **for** *each uncolored neighbor n of g* **do**
8   **if** *there is a suitable color i* **then**
9              $N_i = N_i \cup \{n\}$;
10             $C(n) = i$;
11   **else**
12   Generate a new color c;
13             $N_c = N_c \cup \{n\}$;
14             $C(n) = c$;
15   **end**
16   **end**
17 **end**
18 $\mathcal{N} = \cup_c \{N_c\}$;
19 **return** $\mathcal{N}$;

---

clustering methods contain it to under 200%. At higher split levels, MI computed at the placement layer is not meaningful since attack resilience depends on the distances between the dangling pairs of open pins in the top-most FEOL layer.

If split is done at M1, CCR based on network flow attack is reduced by 3.86x–6.54x when compared with the original design. Random placement can achieve slightly better CCR protection with much higher PPA cost. When splitting at M2, CCR is reduced by 2.73x–3.47x and, at M3, by 1.64x–1.85x. At M4 and above, the reduction range is 1.2x–1.4x on average with a maximum reduction of 1.75x. This implies that at higher levels of split, much care is necessary in placement and routing to defeat proximity based attacks. When compared with the placement perturbation method discussed in Sect. 3.7, the *netlist clustering* method results in 8x lowering of CCR with M4 as the split level.

Gate type (with fanin ignored) based and graph coloring based clustering methods result in 60% area overhead, while gate type clustering with fanin considered to distinguish the type leads to higher area penalty since it results in more clusters. However this is much better than randomized placement. Power overhead is about 50%, which is an improvement over 1.6x over layout randomization. Delay overhead is under 18%, a 5x improvement over randomization.

Netlist clustering has not explicitly considered security-driven placement and routing. Its effectiveness for large benchmarks needs further evaluation.

3.11 Artificial Routing Blockage Insertion

The routing perturbation method, discussed in Sect. 3.8, uses detours and decoys to reroute nets. The concerted wire lifting method, discussed in Sect. 3.9, uses strategic placement of elevating cells to lift wires at selected locations on selected nets. In this section, we will discuss yet another method, proposed by Magaña et al. [28], to directly control routing in order to the proximity and network flow attacks. This method uses two basic steps: *branch insertion* and *blockage insertion*.

3.11.1 Branch Insertion

Branch insertion is a simple technique suggested by Magaña et al. [28] to insert dummy branches on non-critical nets already routed in the design. Each inserted branch starts from a point on a non-critical net and ends on a split layer where it appears as a dummy vpin. The inserted branch will have its end point in the split layer within a selected bounding box centered on the projection of the starting point. While adding these branches, the router will use the available unused routing resources to create artificial congestion on as many nets as possible while ensuring that timing constraints are not violated. Figure 3.13 shows an example of inserted branch and the new vpin created.

Non-trivial branch topologies can be generated using the branch insertion method shown as Algorithm 16. The algorithm begins by eliminating all edges in the routing grid-graph, which are already filled to their capacity (lines 2–6). Next, non-critical nets are examined, and a net with a route fragment on which a branch has not been

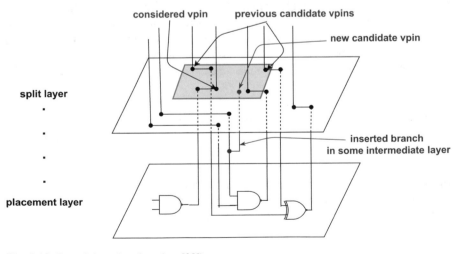

Fig. 3.13 Branch insertion (based on [28])

Algorithm 16: Branch insertion [28]

Input: Split Level l, Layout L, Critical Nets N_c
Output: Modified Layout With Branches L

1 stop = false;
2 **foreach** *routing edge e in the routing grid* **do**
3 | **if** *utilization(e) = capacity(e)* **then**
4 | | Remove e from routing resources;
5 | **end**
6 **end**
7 **while** *stop = false* **do**
8 | stop = true;
9 | **foreach** *net n $\notin N_c$* **do**
10 | | **if** *branch not inserted for route(n)* **then**
11 | | | Randomly select a starting point s on $route(n)$;
12 | | | Compute the maximum length h_{max} for the branch;
13 | | | Create a bounding box B on the split layer l;
14 | | | Determine the shortest routing path p with length h to connect s to B;
15 | | | **if** $h \leq h_{max}$ **then**
16 | | | | Connect s to B via path p;
17 | | | | stop = false;
18 | | | | **foreach** *routing edge e in path p* **do**
19 | | | | | utilization(e) = utilization(e) + 1;
20 | | | | | **if** *utilization(e) = capacity(e)* **then**
21 | | | | | | Remove e from the routing grid;
22 | | | | | **end**
23 | | | | **end**
24 | | | **end**
25 | | **end**
26 | **end**
27 **end**
28 **return** L;

---

inserted is selected (lines 9–10). A random starting point on the route fragment is identified (line 11). Then the maximum length of a branch that can be inserted at that point without violating timing constraints is determined using the slack data and timing estimates. A BB is created on the target split level (line 13). The BB is centered around the same (x,y) coordinates as the starting points (i.e. projection of the starting point on the split layer) and has an area determined by the placement, routing, or routing proximity definitions discussed in Sect. 2.4.1. In the next step, a path is constructed from the starting point to the nearest point in the BB using a shortest path algorithm (line 14). If the shortest path length is within the allowed length, then a branch is created using that path. For all the edges used by the new branch, the edge utilizations are updated and the edges reaching their full capacities are removed from the routing resources (lines 15–24). The process continues as long as new branches can be created on suitable route fragments of non-critical nets.

3.11.2 Blockage Insertion

To defeat proximity attacks, connected pairs of vpins should be spread farther apart. One way to achieve this is to insert artificial blockages in a routed design and reroute the nets. This increases congestion and spreads the vpins apart. Then the blockages are discarded, but the new routes are retained. Based on this technique, Magaña et al. [28, 29] suggested an algorithm, shown as Algorithm 17.

The given split level is first divided into a set of non-overlapping regions called *windows* (line 2). The size of the windows is determined based on benchmark analysis so that a desired number (say, 100) of windows are created. One blockage per window will be inserted. Hence, more windows imply more blockages possibly of smaller sizes. For each window, the available total routing capacity c is determined (lines 6–8). E is the capacity of each routing edge in the global grid-graph. Following this, the size of the blockage $s \times s$ is determined. The blockage should fit within an area equivalent to the estimated number of routing resources available in the window (lines 9–12). The blockage is then inserted at the lower left corner of the window region. It can also be inserted at any other place in the window even if the space is occupied by routed nets since all of the nets will be rerouted after the blockages are inserted. During rerouting, it is likely that nets that previously were routed in the split layer will have to be rerouted using the upper BEOL layers since the lower levels are like to have already been congested. The consequences are an increase in the number of vpins in the split layer and an increase in the distance between the related vpin pairs in the split layer. This will counteract the proximity based attacks.

This basic blockage insertion algorithm may disrupt critical nets that should not be rerouted. Procedure *timing_aware_blockage_insert* shown in Algorithm 17 can be used to handle critical nets. The approach is to turn the critical nets into blockages during rerouting and avoid rerouting those critical nets. For each critical net, the algorithm adjusts the capacities of the edges occupied by the net (lines 17–21). The *blockage_insert* procedure is then called to reroute the non-critical nets after inserting the artificial blockages. After rerouting, the critical nets can be restored, and the artificial blockages should be deleted to obtain the final layout.

3.11.3 Discussion

Magaña et al. [29] reported detailed experiments using five large ISPD-2011 benchmarks. They have used the average size, E[SA], of the search area (in terms of the number of global cells) at each split level (search area is determined by the routing proximity method discussed in Sect. 2.4), the average size, E[LS], of the candidate list, and the average ratio, E[LS/SA], of LS to SA at each split level. E[LS/SA] is a figure of merit (FOM); the higher the value, the more challenging the attack.

When using the basic blockage insertion method with split level set to M4 and using the multi-vpin approach, both E[LS] and FOM increased after insertion.

Algorithm 17: Routing blockage insertion [28]

---

1 blockage_insert *(l, L)*

   Input: Split Level l, Layout L
   Output: Modified Layout With Blockages L

2 Divide layer l into non-overlapping windows W;

3 foreach *window $w \in W$* **do**

4 | Available capacity $c = 0$;

5 | Construct the global routing grid-graph G of w;

6 | **foreach** *edge e in G* **do**

7 | | $c = c + max(0, (E - utilization(e)))$;

8 | **end**

9 | $s = 0$;

10 | **while** $(s + 1)^2 \leq \frac{c}{E}$ **do**

11 | | $s = s + 1$;

12 | **end**

13 | Insert $s \times s$ blockage at the lower-left corner of w in layer l of L;

14 end

15 return L;

---

16 timing_aware_blockage_insert *(l, L, N_c)*

   Input: Split Level l, Layout L, Critical Nets N_c
   Output: Modified Layout With Blockages L

17 foreach *net $n \in N_c$* **do**

18 | **foreach** *global routing edge e in $route(n)$* **do**

19 | | capacity(e) = capacity(e) - 1;

20 | **end**

21 end

22 Delete N_c from L;

23 L = blockage_insert(l, L);

24 return L;

Specifically, the number of vias at and above the split level always increased indicating that more nets are routed in upper levels. Designs continued to be routable even after blockage insertion. Similar trends were observed when 10% of the nets were preserved as being timing critical.

The branch insertion method (without any constraint on wire length) was compared with the blockage insertion method. Branch insertion did result in higher E[LS] and FOM values compared with the original design. However, results showed that the blockage insertion led to better E[LS], while branch insertion resulted in better FOM. This indicates that both are comparable and useful techniques in thwarting proximity based attacks.

It should be noted that, while increase in FOM is a strong indication of defense against proximity based attacks, the branch and blockage insertion methods should be evaluated against stronger attacks such as the network flow attack.

3.12   Netlist Randomization

Most of the defense methods discussed so far have primarily focused on direct modifications to placement and routing. Layout level modifications tend to be time consuming due to the amount of detail to be dealt with at that level and the necessity to run the layout synthesis tools again. Patnaik et al. [30] proposed a method to directly modify the netlist to defend against proximity based attacks.

3.12.1   Objective

In this method, randomly selected pairs of drivers and sinks in the netlist are swapped, and correct connectivity is restored in the BEOL layers with the aid of specially designed cells called *correction cells*. Due to this, truly connected drivers and sinks will be separated by random distances in the FEOL thereby defeating proximity based attacks.

Correction cells work in pairs as shown in Fig. 3.14. D1 and D2 are drive signals and S1 and S2 are sink signals. D1-S1 and D2-S2 are the correct pairs but are swapped with each other. Two correction cells are used in the BEOL layers and are designed as follows: (1) Each correction cell has two input pins C and D and two output pins Y and Z. (2) Arc $C \rightarrow Z$ in each cell is used to realize the erroneous netlist routing. When restoring true connectivity arcs, $C \rightarrow Z$ and $D \rightarrow Y$ are disabled so that only true paths are considered. (3) All pins are set up in higher metal layers to allow lifting and routing of wires in BEOL layers. (4) Correction cells can overlap with standard cells, but a correction cell cannot overlap another correction cell. Legalization should be done accordingly. (5) For proper ECO optimization, the correction cells are set up for load annotation at design time. Note that, corrective rerouting is always done with pairs of correction cells as well as within each correction cell. This further enhances protection against proximity based attacks even when the attacker knows the basic idea of this defense scheme.

Fig. 3.14 Correction cell
(based on [30])

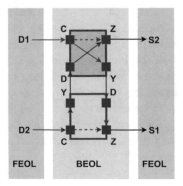

The objective of the method is to swap enough source–sink pairs in the netlist to ensure a desired output error rate and insert appropriate correction cells before layout synthesis.

3.12.2 Algorithm

The netlist randomization method is shown as Algorithm 18. Randomly selected pairs of drivers and sinks in the netlist are swapped while ensuring that the swap does not introduce a combinational cycle in the netlist (lines 3–5). Enough nets are selected to ensure that the OER approaches 100%, that is, the modified netlist introduces some output errors for any input (line 6). Correction cells are inserted to facilitate wire lifting (line 7). The erroneous netlist is now placed and routed after setting "don't touch" flags on the swapped nodes to avoid logic optimization (lines 8–9). Global optimization for timing, power, congestion, etc. can be performed as usual. Finally, the design is rerouted in the ECO (engineering change order) mode to implement lifting of the swapped nets. Their correct connectivity is restored in the BEOL layers using the correction cells. After rerouting, post-route optimization can be performed (lines 9–10). This process is repeated as long as the PPA budget is satisfied. At the end, the correction cells are removed and the FEOL and BEOL layouts are produced.

Algorithm 18: Netlist randomization [30]

Input: Netlist N, PPA Budget B
Output: Protected FEOL Layout, BEOL Layout
1 **while** *PPA budget B allows additional randomization* **do**
2   **repeat**
3   Randomly select two suitable nets n_1 and n_2;
4   Make a note of the connectivity information for n_1 and n_2;
5   Swap the connectivity (drivers and sinks) of n_1 and n_2;
6   **until** *output error rate (OER) approaches 100 %*;
7   Embed correction cells;
8   Set "don't touch" flags for the swapped drivers and sinks;
9   Place and route and lift erroneous nets;
10   Reroute in BEOL using correction cells, legalize, post-route optimize in ECO mode;
11 **end**
12 Remove correction cells;
13 **return** FEOL Layout, BEOL Layout

3.12.3 Discussion

Patnaik et al. [30] have evaluated the method using ISCAS-85 and IBM superblue benchmarks. Up to 10 metal layers were used. Netlist randomization increased distances between connected gates by a factor of about 20x on average (with standard deviation of about 6x) when compared with either the original layouts or a naive lifting method. Majority of wiring moved to upper metal layers with increased vias/vpins in higher layers. While naive lifting or placement perturbation methods provided no or weak protection, netlist randomization protects routing and defends against both placement based and routing based attacks. If M5 is taken as the split layer, vias between M5 and M6 increased by 30.65%, which defends against the crouting attack [28] discussed in Sect. 3.11.

The network flow attack and the crouting attack were used to verify attack resilience. With placement perturbation defense (Sect. 3.7), the network flow attack resulted in average CCR, OER, and HD values of about 92%, 85%, and 15%, respectively. Various clustering methods [26] discussed in Sect. 3.10 help reduce CCR to 57%–66%. Pin swapping method [6], discussed in Sect. 3.4, leads to 88% CCR and 33% HD. Routing perturbation [21], discussed in Sect. 3.8, leads to 72% CCR, 100% OER, and 29% HD. Netlist randomization method reduces CCR to 0% while yielding OER of nearly 100% and HD of about 40% on average. These numbers indicate the superiority of netlist randomization in protecting against both placement and routing hints. This came at the expense of 11.5% increase in power and 10% increase in delay for the relatively small ISCAS-85 benchmarks. Respective overheads were 3.5% and 2.7% for the large superblue benchmarks.

Effectiveness of netlist randomization shows that defense methods are easier to plan at a higher level in the design process and tend to be more effective in increasing security with only modest PPA and design time overheads.

3.13 Summary

In this chapter, we have discussed several defense methods to thwart both greedy and clever proximity based attacks. Table 3.1 shows a summary of these methods. While these defenses were tested against the proximity and network flow attacks, the same strategies should be effective against the other design constraint based attacks discussed in Chap. 2.

Table 3.1 Summary of defenses against design constraint based attacks

Sl.	Defense method	Year	Attacks thwarted	Benchmarks	Metrics
1	Pin swapping	2013	Proximity	ISCAS-85	HD, AC
2	Secure min-cut bipartitioning	2015	Proximity	ISCAS-85, ITC-99	HD, AC
3	Secure multiway partitioning	2018	Proximity	ISCAS-85, ITC-99	HD, AC
4	Placement perturbation	2018	Proximity	ISCAS-85, ITC-99	OER CCR
5	Routing perturbation	2017	Network flow	ISCAS-85, ITC-99	ICR, OER, HD
6	Concerted wire lifting	2018	Network flow	ISCAS-85, ISPD-11	PNR
7	Netlist clustering	2017	Network flow	ISCAS-85, MCNC	MI, CCR
8	Artificial routing blockage insertion	2017	Routing proximity	ISPD-11	E[LS], FOM
9	Netlist randomization	2018	Network flow, crouting proximity	ISCAS-85, ISPD-11	CCR, OER, HD

References

1. M. Jagasivamani, P. Gadfort, M. Sika, M. Bajura, M. Fritze, Split-fabrication obfuscation: Metrics and techniques, in *Proceedings of the 2014 IEEE International Symposium on Hardware-Oriented Security and Trust, HOST 2014* (2014), pp. 7–12
2. C.T.O. Otero, J. Tse, R. Karmazin, B. Hill, R. Manohar, Automatic obfuscated cell layout for trusted split-foundry design, in *Proceedings of the 2015 IEEE International Symposium on Hardware-Oriented Security and Trust, HOST 2015* (Institute of Electrical and Electronics Engineers Inc., Piscataway, 2015), pp. 56–61
3. Y. Xie, C. Bao, A. Srivastava, 3D/2.5D IC-based obfuscation, in *Hardware Protection through Obfuscation* (Springer International Publishing, Berlin, 2017), pp. 291–314
4. Y. Alkabani, Hardware security and split fabrication. Int. Desig. Test Workshop **0**, 59–64 (2016)
5. M.A. Masoud, Y. Alkabani, M.W. El-Kharashi, Obfuscation of digital systems using iso-morphic cells and split fabrication, in *Proceedings – 2018 13th International Conference on Computer Engineering and Systems, ICCES 2018* (Institute of Electrical and Electronics Engineers Inc., Piscataway, 2019), pp. 488–493
6. J. Rajendran, O. Sinanoglu, R. Karri, Is split manufacturing secure? in *Proceedings – Design, Automation and Test in Europe* (2013), pp. 1259–1264
7. J. Rajendran, O. Sinanoglu, R. Karri, System, method and computer-accessible medium for providing secure split manufacturing. US Patent, 10423749 B2 (2019)
8. J. Rajendran, Y. Pino, O. Sinanoglu, R. Karri, Logic encryption: a fault analysis perspective, in *Proceedings – Design, Automation and Test in Europe* (2012), pp. 953–958
9. J. Rajendran, H. Zhang, C. Zhang, G.S. Rose, Y. Pino, O. Sinanoglu, R. Karri, Fault analysis-based logic encryption. IEEE Trans. Comput. **64**(2), 410–424 (2015)
10. M. Bushnell, V.D. Agrawal, *Essentials of Electronic Testing for Digital, Memory and Mixed-Signal VLSI Circuits* (Springer, Berlin, 2002)

11. Y. Xie, C. Bao, A. Srivastava, Security-aware design flow for 2.5D IC technology, in *TrustED 2015 – Proceedings of the 5th International Workshop on Trustworthy Embedded Devices, co-located with CCS 2015* (2015), pp. 31–38
12. C.M. Fiduccia, R.M. Mattheyses, A linear-time heuristic for improving network partitions, in *Proceedings – Design Automation Conference* (Institute of Electrical and Electronics Engineers Inc., Piscataway, 1982), pp. 175–181
13. L.T. Wang, Y.W. Chang, K.T. Cheng, *Electronic Design Automation: Synthesis, Verification, and Test* (Morgan Kaufmann, Burlington, 2009)
14. A. Waksman, M. Suozzo, S. Sethumadhavan, FANCI: identification of stealthy malicious logic using Boolean functional analysis, in *Proceedings of the ACM Conference on Computer and Communications Security* (ACM Press, New York, 2013), pp. 697–708
15. Y.C. Chang, Y.W. Chang, G.M. Wu, S.W. Wu, B*-trees: a new representation for non-slicing floorplans, in *Proceedings – Design Automation Conference* (2000), pp. 458–463
16. Y.K. Ho, Y.W. Chang, Multiple chip planning for chip-interposer codesign, in *Proceedings – Design Automation Conference* (2013), pp. 1–6
17. S. Chen, R. Vemuri, Improving the security of split manufacturing using a novel BEOL signal selection method, in *Proceedings of the ACM Great Lakes Symposium on VLSI, GLSVLSI* (2018), pp. 135–140
18. Ç. Ümitand, A. Cevdet, PaToH (partitioning tool for hypergraphs), in *Encyclopedia of Parallel Computing* (Springer, Berlin, 2011), pp. 1479–1487
19. P. Spindler, U. Schlichtmann, F.M. Johannes, Kraftwerk2 – a fast force-directed quadratic placement approach using an accurate net model. IEEE Trans. Comput. Aided Design Integ. Circuits Syst. **27**(8), 1398–1411 (2008)
20. Y. Wang, P. Chen, J. Hu, G. Li, J. Rajendran, The cat and mouse in split manufacturing. IEEE Trans. Very Large Scale Integ. (VLSI) Syst. **26**(5), 805–817 (2018)
21. Y. Wang, P. Chen, J. Hu, J. Rajendran, Routing perturbation for enhanced security in split manufacturing, in *Proceedings of the Asia and South Pacific Design Automation Conference, ASP-DAC* (Institute of Electrical and Electronics Engineers Inc., Piscataway, 2017), pp. 605–610
22. L.H. Goldstein, E.L. Thigpen, SCOAP: Sandia controllability/observability analysis program, in *Proceedings – Design Automation Conference* (1980), pp. 190–196
23. Y. Wang, P. Chen, J. Hu, J.J. Rajendran, The cat and mouse in split manufacturing, in *Proceedings – Design Automation Conference*, vol. 05 (2016), pp. 1–6
24. S. Patnaik, J. Knechtel, M. Ashraf, O. Sinanoglu, Concerted wire lifting: enabling secure and cost-effective split manufacturing, in *Proceedings of the Asia and South Pacific Design Automation Conference* (2018), pp. 251–258
25. N. DfX Lab, Design-for-Trust Technique (2017) [Online]. Available: https://sites.nyuad.nyu.edu/dfx/research-topics/design-for-trust-split-manufacturing/
26. A. Sengupta, S. Patnaik, J. Knechtel, M. Ashraf, S. Garg, O. Sinanoglu, Rethinking split manufacturing: an information-theoretic approach with secure layout techniques, in *IEEE/ACM International Conference on Computer-Aided Design, Digest of Technical Papers, ICCAD*, vol. 2017 (2017), pp. 329–336
27. R. Lewis, *A Guide to Graph Colouring* (Springer, Berlin, 2016)
28. J. Magana, D. Shi, J. Melchert, A. Davoodi, Are proximity attacks a threat to the security of split manufacturing of integrated circuits? IEEE Trans. Very Large Scale Integ. (VLSI) Syst. **25**(12), 3406–3419 (2017)
29. J. Magaña, D. Shi, A. Davoodi, Are proximity attacks a threat to the security of split manufacturing of integrated circuits? in *IEEE/ACM International Conference on Computer-Aided Design, Digest of Technical Papers, ICCAD*, vol. 07 (2016), pp. 1–7
30. S. Patnaik, M. Ashraf, J. Knechtel, O. Sinanoglu, Raise your game for split manufacturing: restoring the true functionality through BEOL, in *55th ACM/ESDA/IEEE Design Automation Conference* (IEEE, New York, 2018), pp. 1–6

Chapter 4
Satisfiability Based Attacks

Abstract In this chapter, we discuss an oracle-guided attack called the satisfiability (SAT) attack against split manufactured ICs. In the design constraint based attacks, usually an attacker cannot recover all of the BEOL signals correctly. In contrast, the SAT attack can reveal 100% of the hidden signals. We begin the chapter by discussing three background topics: satisfiability checking, logic locking through key gates, and the SAT attack against logic locking. We discuss a method for netlist to layout mapping using satisfiability in the context of trojan insertion attacks. We then introduce the application of the SAT attack for reverse engineering a split design and discuss in detail the methods for transforming the BEOL signal recovery problem into the problem of logic decryption. We discuss methods for elimination of cycles from the equivalent encrypted circuit corresponding to incorrect BEOL recoveries. We show how to extend the SAT attack to split manufactured sequential circuits. We discuss how to incorporate proximity constraints into a SAT attack to reduce the attack time and increase the attack capacity. Finally, we discuss a method to group the BEOL signals using a satisfiability-modulo theories (SMT) solver to further reduce the size of the SAT problem and increase attack efficiency.

In Chap. 2, we have discussed several attacks which exploit inferences derived from the FEOL layout and common knowledge of the design automation algorithms to formulate a set of constraints to identify the correct BEOL nets. Success rate of such attacks is heavily dependent on the correctness of the constraints and the order in which conflicting constraints are applied. In this chapter, we will discuss alternative formulations of the reverse engineering and trojan insertion attacks using satisfiability checking. These attacks usually guarantee 100% attack success at the expense of significant computational resources.

© The Author(s), under exclusive license to Springer Nature Switzerland AG 2021   105
R. Vemuri, S. Chen, *Split Manufacturing of Integrated Circuits for Hardware Security and Trust*, https://doi.org/10.1007/978-3-030-73445-9_4

4.1   Background

4.1.1   Satisfiability Checking

Given a Boolean function f of n Boolean variables, f is *satisfiable* if there exists an assignment of values to the n variables that would satisfy the function, that is, evaluate it to true. A satisfying variable assignment is called a model. If there is no satisfying assignment, then the function is unsatisfiable. The satisfiability checking (SAT) problem [1] is to determine a model for a Boolean function if one exists or show that the function is unsatisfiable. A SAT solver is a computational procedure to solve SAT problem instances. Satisfiability checking is an extensively studied problem and continues to be an active area of research [2–4].

Problem instances for SAT solvers are usually specified in the form of *CNF (Conjunctive Normal Form)* expressions. A CNF expression is a conjunction of *clauses* where a clause is a disjunction of *literals*. A literal is a positive or negative Boolean variable. For example, the following are CNF expressions:

$$f(a, b, c) = (a \vee b') \wedge (b \vee c') \wedge (c \vee a') \qquad (4.1)$$

$$g(a, b, c) = (a \vee c') \wedge (a' \vee b') \wedge (c) \wedge (b \vee c') \qquad (4.2)$$

f is satisfiable. Setting all three variables to *true* is one possible solution for f. Setting all three variables to *false* is another possible solution for f. A SAT solver can return any satisfying assignment. g is not satisfiable; a SAT solver should return UNSAT for g. Note that the CNF expressions are also called product-of-sums (POS) expressions. f and g can also be written using an alternative notation as,

$$f(a, b, c) = (a + b').(b + c').(c + a') \qquad (4.3)$$

$$g(a, b, c) = (a + c').(a' + b').(c).(b + c') \qquad (4.4)$$

Satisfiability solving is equivalent to determining if it is possible to make the output of a logic circuit high and, if so, producing any one input vector to accomplish it. The function of any logic circuit can be expressed in the CNF form and submitted to a SAT solver.

SAT problem is a fundamental NP-complete problem [5]. Heuristics for SAT solving have been steadily improving in the past two decades [6]. Vast majority of modern SAT solvers are based on the classic DPLL procedure [7] and incorporate some form of conflict-driven clause learning (CDCL) [8]. In addition, efficient BCP (Boolean Constraint Propagation) heuristics and conflict resolution strategies are incorporated for fast convergence. SAT solving is central to several formal verification techniques including equivalence checking [9] and bounded model checking [10]. Both of these tools are widely used to establish basic functional correctness which is necessary but often not sufficient for trust assurance. Many trust assurance problems can be modeled and solved as SAT problems and several attack

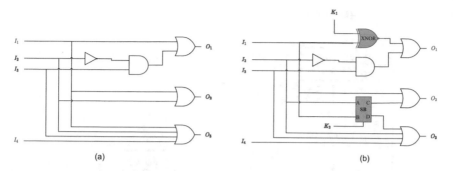

Fig. 4.1 Logic encryption example. (**a**) Example logic circuit, (**b**) Encrypted logic circuit

and defense methods use SAT algorithms. Examples of security and trust problems that make use of SAT solvers include *logic encryption*, circuit camouflaging, split manufacturing, trojan detection, side-channel defense, reverse engineering, and watermarking.

4.1.2 Logic Encryption

Logic encryption (also called logic locking or logic obfuscation) [11] is a design-for-trust method to prevent counterfeiting, overproduction, reverse engineering, and unauthorized use of an integrated circuit. In this method, the function of the circuit is corrupted by the introduction of new gates called *key gates* controlled by new inputs called *key inputs*. When correct key values are applied, the circuit works as expected. However, if incorrect key values are applied, incorrect outputs are produced for one or more input vectors. For example, Fig. 4.1a shows a logic circuit and Fig. 4.1b shows an encrypted version of the same logic circuit. K_1 and K_2 are the newly introduced key inputs. SB is a switch box ($C = AK' + BK, D = BK' + AK$). Correct key values are $K_1 = 1$ and $K_2 = 0$. Incorrect key values may corrupt one or more primary outputs depending on the primary input values.

A design process with logic encryption is shown in Fig. 4.2a. During the encryption step, the logic design is encrypted by including the key gates and key inputs. Layout of the encrypted design is sent to an untrusted foundry. The correct key values are provided to a legitimate buyer of the IC, such as a system integrator. The system integrator stores the correct key values in a tamper-proof memory and packages it along with the IC such that the key bits are secured from an attacker as illustrated in Fig. 4.2b.

Logic encryption was first proposed by Roy et al. in 2008 [12] based on the idea of inserting XOR or XNOR gates at random locations in a netlist. Several other encryption methods were proposed in the next few years [13–16]. Attacks against logic encrypted circuits began to appear from 2012. Rajendran et al. [14] proposed

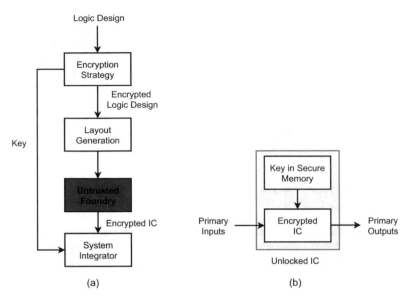

Fig. 4.2 Logic encryption process. (**a**) IC design, fabrication, and usage, (**b**) Integration of the key

an attack based on the fault sensitization methods used in IC testing and further proposed a strong logic locking method [17] that is resilient against the sensitization attack.

4.1.3 Satisfiability (SAT) Attack Against Logic Encryption

In 2015, Subramanyan et al. [18] proposed an attack using *satisfiability checking* to recover the key values assuming that the attacker has access to the encrypted logic netlist and an unlocked IC or its simulation model which can be used to make a limited number of input–output observations. A SAT solver guides the choice of the input vectors for which observations are needed.

The SAT solver uses the miter circuit shown in Fig. 4.3 to generate these input vectors. C_A and C_B are two copies of the encrypted netlist using the same input vector I but different key vectors K_A and K_B. The output vectors are O_A and O_B respectively. A comparator is used to compare the output vectors and produce a *diff* signal if the two vectors differ in at least one bit. This miter circuit is represented as an equivalent CNF formula and submitted to a SAT solver which is asked to satisfy the *diff* signal if possible. If so, the SAT solver generates instances i, k_A, k_B, o_A, and o_B such that $o_A \neq o_B$ and $k_A \neq k_B$. The input vector instance i that can satisfy *diff* is called a *distinguishing input pattern* (DIP); it can distinguish between k_A and k_B by producing two different output vectors. The oracle (unlocked IC or simulation model) is then used to obtain the true output instance o when the DIP i is applied.

Fig. 4.3 Miter circuit used
by the satisfiability attack

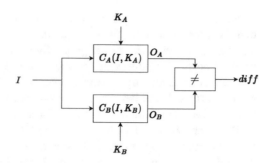

Algorithm 19: Satisfiability (SAT) attack [18]

Input: Netlist N, Oracle $IC(I)$
Output: Correct Key Vector Instance k
1 Prepare CNF $C(I, K, O)$ for N and make two copies $C(I, K_A, O_A)$ and $C(I, K_B, O_B)$;
2 $F = C(I, K_A, O_A) \wedge C(I, K_B, O_B)$;
3 **while** $SAT(F \wedge (O_A \neq O_B))$ **do**
4 $\quad$ $i =$ assignment to I determined by $SAT(F \wedge (O_A \neq O_B))$;
5 $\quad$ $o = IC(i)$;
6 $\quad$ $F = F \wedge C(i, K_A, o) \wedge C(i, K_B, o)$;
7 **end**
8 $k =$ assignment to K_A determined by $SAT(F)$;
9 **return** k;

This observation is added in the form of constraints $C(i, K_A) = o$ and $C(i, K_B) = o$
to the SAT problem and the process is repeated until no further DIPs exist at which
point all the incorrect keys are eliminated and the correct key is found. The SAT
attack process is shown as Algorithm 19.

The power of the SAT attacks is due to the way incorrect key values are elimi-
nated. Two key vector instances k_A and k_B are *equivalent* provided $\forall i, C(i, k_A) = C(i, k_B)$. Each DIP helps eliminate at least one equivalence class of incorrect key
vectors. By discovering and eliminating entire sets of key vectors in each iteration,
the SAT attack quickly converges on the class of correct key vectors and returns a
member of that class.

Subramanyan et al. [18] demonstrated that SAT attack can defeat all of the logic
encryption methods existing at that time, including [12–16]. Soon thereafter, SAT
attack resilient encryption methods began to appear. These include the SARLock
[19], Anti-SAT [20], SFLL [21], and others.

4.1.4   Cyclic Logic Locking and CycSAT Attack

Shamsi et al. [22] proposed the *cyclic logic locking* method to thwart the SAT attack.
This method systematically introduces key-controlled gates and redundant wires
into the logic netlist such that incorrect key values activate cyclic paths through
the redundant wires. In addition, to ensure that these redundant wires are not easily

identified and removed, the method inserts multiple entry points into each cycle so that the forward and backward edges in a cycle cannot be uniquely identified by a simple traversal of the netlist graph. Shamsi et al. reasoned that the SAT solvers accept Boolean formulas representing combinational logic circuits and, hence, the SAT attack is inherently an attack against encrypted combinational logic circuits without cycles. If the key logic is introduced such that incorrect key values cause cyclic paths in the circuit, then the SAT attack cannot be used.

However, Zhou et al. [23] demonstrated that when the SAT attack is used against cyclic encryption, one of the following three outcomes is possible depending upon the order in which the DIPs are produced by the SAT solver: (1) The attack terminates and recovers the correct key values which result in the correct acyclic combinational circuit. (2) The attack terminates but the recovered key values result in an incorrect, cyclic circuit. (3) The attack goes into an infinite loop, repeatedly producing the same DIPs. Zhou et al. proposed adding certain "No-Cycle" (NC) constraints to the SAT attack such that the SAT attack can recover the original acyclic circuit. The enhanced attack is called the *CycSAT attack*. The *NC constraints* on the key values capture the conditions under which every cyclic path in the circuit is broken. These conditions are expressed as CNF formulas and are added to the circuit model. We will discuss the CycSAT attack method in the context of adapting it to attack SM in Sect. 4.3.

4.2 Satisfiability Based Layout Recognition

Recall that the layout recognition attacks against SM are concerned with identifying a given circuit structure in the FEOL layout. The goal of the attacker is to insert a trojan at the matching location. Hence these attacks are also called *trojan insertion attacks*. Imeson et al. [24] formulated layout recognition as a graph matching problem.

Given a netlist graph G which represents circuit netlist and a layout graph H which represents a netlist extracted from the FEOL layout, the attacker is attempting to construct a mapping between the nodes in G and the nodes in H. Formally, the attacker's goal is to find if there is subgraph of G which is isomorphic to H, that is, whether G is *subgraph isomorphic* to H. Imeson et al. formulated the problem of subgraph isomorphism as a satisfiability problem as follows:

Let $G = (V_g, E_g)$ be the netlist graph and $H = (V_h, E_h)$ be the layout graph. Boolean variables b_{ij}, $1 \leq i \leq |V_h|$, $1 \leq j \leq |V_g|$, are introduced such that b_{ij} is true if and only if vertex $i \in V_h$ is mapped to vertex $j \in V_g$. Boolean formulas constraining b_{ij} are constructed as follows:

1. Each vertex in G maps to only one vertex in H:

$$F_1 = \prod_{i \in V_h} \sum_{j \in V_g} \left(b_{ij} \prod_{k \in V_g, k \neq j} b'_{ik} \right) \qquad (4.5)$$

2. Each vertex in H maps to only one vertex in G:

$$F_2 = \prod_{j \in V_g} \sum_{i \in V_h} \left(b_{ij} \prod_{k \in V_h, k \neq i} b'_{kj} \right) \qquad (4.6)$$

3. Each edge in E_h maps to an edge in E_g. Let the source and destination nodes of any edge e be denoted by $src(e)$ and $dst(e)$.

$$F_3 = \prod_{k \in E_h} \sum_{l \in E_g} \left(b_{src(k)src(l)} \wedge b_{dst(k)dst(l)} \right) \qquad (4.7)$$

Finally, the conjunction of the three formulas captures all the constraints on the mapping

$$F = F_1 \wedge F_2 \wedge F_3 \qquad (4.8)$$

F is satisfiable if and only G is subgraph isomorphic to H; the satisfying model of F, that is the set of assignments to the b variables, yields the subgraph of G which is isomorphic to H. The mapping method is summarized as Algorithm 20. SAT solvers return one solution. However, by adding the previous solutions as constraints, a new solution, if one exists, can be produced.

While they have not used this method to mount an attack, Imeson et al. used it to determine the k-security level of a netlist from which some wires are lifted and then defined a greedy heuristic for wire lifting to ensure k-security. This wire lifting procedure will be discussed in Sect. 5.1.

Algorithm 20: Layout recognition [24]

Input: Netlist Graph $G = (V_g, E_g)$, Layout Graph $H = (V_h, E_h)$
Output: Bijective Mapping Between V_g and V_h
1 **for** *each pair of vertices $i \in V_h$ and $j \in V_g$* **do**
2 | create a Boolean variable b_{ij};
3 **end**
4 Create Boolean formula F_1 per Eq. 4.5;
5 Create Boolean formula F_2 per Eq. 4.6;
6 Create Boolean formula F_3 per Eq. 4.7;
7 $F = F_1 \wedge F_2 \wedge F_3$;
8 B = Satisfiability_Solver(F);
9 **return** (i, j) pairs for which b_{ij} are set to true in B;

4.3 SAT Attack Based Reverse Engineering

Chen et al. [25] proposed a satisfiability based, oracle-guided reverse engineering attack to recover the missing BEOL nets given the FEOL netlist and a working IC or a functional model of the IC. This attack in turn uses the SAT attack, discussed in Sect. 4.1.3), proposed by Subramanyan et al. [18] to decrypt a logic encrypted circuit. A working IC, possibly one produced using an older process, could be purchased by the attacker in the open market. The attacker might have stolen a simulation model from the design house or might have access to a functional equivalent. In all these cases, the attacker has an *oracle* which is sufficient to mount the SAT attack.

The problem is formulated as follows: Let $G = (V, E)$ be a directed acyclic hypergraph representation of a combinational logic circuit. Vertices v_i, $1 \le i \le |V|$ represent logic gates and hyperedges e_j, $1 \le j \le |E|$ represent signal nets. Nodes with no incoming (outgoing) edges represent the primary input (output) terminals. Let $H \subseteq E$ be the subset of nets selected to be assigned to the "hidden" BEOL layers. Remaining circuit $(V, E - H)$ is assigned to the FEOL layers. The attacker already knows $(V, E - H)$. For a certain input pattern I_x, the correct output from a packaged IC is O_x. The problem is to recover H correctly by using several strategically selected (I_x, O_x) pairs that the attacker obtained using the available oracle.

Removal of BEOL nets H induces non-primary input and output nodes (with no incoming and outgoing edges, respectively) in the FEOL circuit $(V, E - H)$. These non-primary input (output) nodes of the FEOL circuit are also referred to as the output (input) nodes or terminals of BEOL nets.

4.3.1 Modeling BEOL Recovery as Logic Decryption

The problem of recovering the BEOL nets is formulated as a logic decryption problem as follows. The missing H signals, when reintroduced, should connect certain output terminals to certain input terminals in the FEOL circuit. These terminals are internal to the original circuit but available as external terminals of the FEOL circuit. A key-controlled network is introduced to model the connectivity of these output terminals to the input terminals. Key inputs represent the uncertainty about output–input mapping. Correct key values will program the network correctly to recover the missing BEOL connections and hence the correct circuit. A key-controlled *crossbar network*, such as the Benes network [26] would be a good candidate interconnect network to model this connectivity. This network can establish all possible connections between inputs and outputs based on the key values. However, the number of the gates to represent this network is too large leading to too many clauses for the SAT solver. An interconnect network which

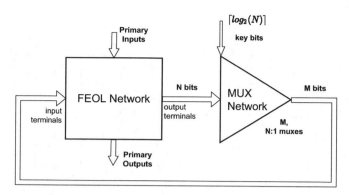

Fig. 4.4   General multiplexor network

results in fewer SAT clauses is preferred. A multiplexer network can be used to reduce the complexity at the expense of more key bits.

Figure 4.4 shows a general MUX network structure that allows any FEOL output terminal to drive any FEOL input terminal. The network consists of M, N-to-1 multiplexors, one for each FEOL input terminal. The N-to-1 multiplexer (MUX) allows its output to be connected to any of its inputs by controlling the select signals. A N-to-1 multiplexer M has inputs $i_1, i_2, \ldots, i_N$ and $n = \lceil log_2(N) \rceil$ select signals which are also called the *key bits*. The key bits are denoted by a vector of Boolean variables $K_M = \langle k_1, k_2, \ldots k_n \rangle$. Each input signal i of the MUX is associated with a unique binary valued assignment $B_M^i = \langle b_1, b_2, \ldots, b_n \rangle$ to the key bits such that when $K_M = B_M^i$, the output of the MUX is connected to i. For the missing BEOL nets $e \in H$, all the source nodes of e constitute the input terminals of the MUX network and the target nodes of e constitute the output terminals of the MUX network. This network structure contains more connection combinations than the crossbar network since it allows for a BEOL input terminal to be mapped to multiple BEOL output terminals. However this is not an issue since the SAT solver will automatically preclude the key combinations that lead to functionally incorrect circuits. The key bits of all the multiplexers need to be determined by the SAT attack subject to the constraint that each BEOL output terminal must be driven by a unique BEOL input terminal.

If the attacker is able to correctly separate the FEOL netlist into multiple partitions assuming that the design process was based on min-cut or similar partitioning as suggested in Sect. 2.3 [27], then the mux network can be simplified to the one shown in Fig. 4.5a. This network can also model the split fabrication using 2.5D or 3D integration in which the circuit is split into two (or more) chips and the interconnections between the two chips are hidden in a trusted interconnect layer. Xie et al. [28] and Wang et al. [29] also proposed similar MUX networks for modeling the BEOL interconnect recovery problem as a SAT attack problem for encrypted logic circuits.

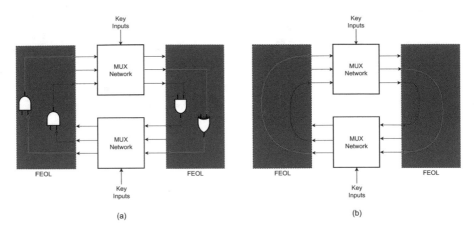

Fig. 4.5 MUX network for two partitions (based on [25]). (**a**) Design with MUX network, (**b**) Simplified representation

4.3.2 Elimination of Cyclic Paths

Introduction of the MUX network leads to potential cyclic paths in the circuit. While numerous cyclic paths are physically present, none of those paths will be activated with the correct MUX key bits. However, there is a possibility to generate many *combinational cycles* during the attack process corresponding to incorrect key guesses and these cycles can confuse the *SAT solver* and thwart the SAT attack. Hence, constraints on the key values that would avoid activating the cyclic paths are generated. This requires identification of all the cycles in order to find the complete set of key constraints for all combinational cycles. However, if each cycle through each gate is analyzed similar to CycSAT [23], the *no-cycle* (NC) conditions will not necessarily cover all possible combinational cycles in polynomial time as discussed in detail by Roshanisefat et al. [30]. CycSAT application context dealt with cyclic obfuscations which tend to introduce far fewer and less complex cycles than the MUX networks.

The extracted FEOL circuit would be cycle free if the original circuit is cycle free. So the cycles, if any, would be generated by the MUX networks introduced earlier. In order to locate all the cycles, the complete circuit needs to be analyzed using *depth first search* [31]. To increase the efficiency of this search, a simplified *MUX graph G'* is produced. The idea is illustrated with this example: In Fig. 4.5a, a circuit with MUX based interconnect networks is shown. In Fig. 4.5b, the internal gates are removed and direct edges in place of directed paths from input to output terminals in each FEOL partition are introduced. The attacker only needs to check for cycles in this simplified circuit.

A simplified graph called the *MUX graph, G'*, is constructed as follows: each FEOL terminal is represented by a node in G'. If there is an edge (i.e. a path before the above simplification) from input terminal i to output terminal o in the FEOL

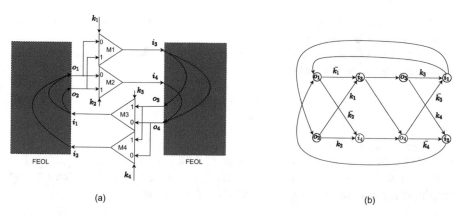

Fig. 4.6 Multiplexor graph used for cycle generation (based on [25]), (**a**) Simplified MUX network, (**b**) MUX graph

netlist, then a directed edge from node i to node o is introduced in G'. If an input terminal i can be driven by an output terminal o through a MUX controlled by a key bit k, then an edge from node o to node i is introduced in G' and the edge is labeled with the enabling key value. Figure 4.6a shows circuit with the simplified FEOL graphs and Fig. 4.6b shows the corresponding MUX graph. The MUX graph G' represents the multiplexers and the combinational paths among the multiplexers through the FEOL circuit. None of the circuit gates is directly represented. However, the following observation can be made: every path between two terminals in the simplified netlist has a corresponding path between the two nodes corresponding to the two terminals. Hence, any cyclic path in the netlist has a corresponding cycle in G'. Clearly, G' is sufficient to determine all the cycles introduced by the MUX network in the FEOL circuit. The depth-first search algorithm described in [31] is used to generate all the cycles in the MUX graph.

Once all possible combinational cycles are traced, constraints are generated for all the MUXes in each combinational cycle. Suppose a cyclic path contains multiplexers $M_1, M_2, \ldots M_m$ controlled by keys $K_1, K_2, \ldots K_m$ respectively where each K is a vector of Boolean variables. Intuitively, at least one of these MUXes should be "disabled" in order to disable the cyclic path. That is, at least one of $K_i, 1 \leq i \leq m$, must be set to a binary vector such that the cyclic path is *disabled* through M_i. Suppose key K_i is the Boolean variable vector $K_i = \langle k_{i_1}, k_{i_2}, \ldots k_{i_{|K_i|}} \rangle$ and that the variables must assume the binary values $B_i = \langle b_{i_1}, b_{i_2}, \ldots b_{i_{|K_i|}} \rangle$, $b_{i_j} \in \{0, 1\}$ in order to enable the correct path through the multiplexer M_i. This cyclic path would be *enabled* if every MUX M_i in the path is enabled correctly. This condition can be expressed as,

$$\bigwedge_{i \in \{1, m\}} (K_i = B_i) \tag{4.9}$$

Referring to Fig. 4.6a, the cyclic path $o_1 \rightarrow i_3 \rightarrow o_3 \rightarrow i_1 \rightarrow o_1$ would be enabled if $(k_1 = 0) \wedge (k_3 = 1)$.

For M_i to be disabled, at least one of these keys should be incorrectly set. This can be expressed as a logical constraint of the form,

$$\bigvee_{i \in \{1,m\}} (K_i \neq B_i) \tag{4.10}$$

In the above example, the cyclic path would be disabled provided $(k_1 \neq 0) \vee (k_3 \neq 1)$. Since each key K_i consists of several bits $K_i = \langle k_{i_1}, k_{i_2}, \ldots k_{i_{|K_i|}} \rangle$, for $K_i \neq B_i$, at least one of these bits must be incorrectly set. Hence, the logical condition for disabling the cyclic path can be expressed as,

$$\bigvee_{i \in \{1,m\}, j \in \{1,|K_i|\}} (k_{i_j} \oplus b_{i_j}) \tag{4.11}$$

For the example above, assuming that $K_1 = \langle k_1 \rangle$ and $K_3 = \langle k_3 \rangle$, the following SAT constraint clause can be generated to disable the cycle: $k_1 \vee \bar{k}_3$. For each cycle in the MUX graph G', a SAT constraint clause of the form shown in Eq. 4.11 should be generated to disable that cycle.

Table 4.1 shows the list all combinational cycles for the MUX graph shown in Fig. 4.6 and the corresponding key constraints to disable these cycles. These constraints are called *No-Cycle* (NC) constraints.

The cycle constraint generator can enumerate all the cyclic paths in a circuit and generate the corresponding NC constraints in the form of CNF clauses. However, for large circuits, the number of these clauses can exceed the capacity of state-of-the-art SAT solvers. Hence, these constraints are further reduced based on the following observation: if P_1 and P_2 are two cyclic paths in the graph G' such that every edge in $P1$ is also an edge in $P2$, then disabling the $P1$ path in the circuit implies that $P2$ is also disabled. Hence, it is sufficient to include the constraints for $P1$ and omit those for $P2$. That is, if a cycle involving a specific sequence of multiplexers $M_x, \ldots M_y$ with a specific set of key values ($K_x = B_x, \ldots, K_y = B_y$) is identified and the corresponding constraint is generated, then there is no need to introduce a constraint for another cyclic path which contains the same multiplexers with the same key values for each multiplexer.

Given two NC constraint clauses C_1 and C_2, C_2 is redundant provided all the literals in C_1 also appear in C_2. Consider the following example set of cycle constraints from rows 1, 2, 6, and 7 in Table 4.1:

$$k_1 \vee \bar{k}_3$$

$$k_1 \vee \bar{k}_3 \vee \bar{k}_2 \vee k_4$$

$$k_2 \vee k_3$$

$$k_2 \vee k_3 \vee \bar{k}_1 \vee \bar{k}_4$$

Table 4.1 Combinational cycles and no-cycle constraints

	Combinational cycle	No-cycle constraint	Redundant?
1	$o_1 \rightarrow i_3 \rightarrow o_3 \rightarrow i_1 \rightarrow o_1$	$k_1 \vee \bar{k}_3$	No
2	$o_1 \rightarrow i_3 \rightarrow o_3 \rightarrow i_1 \rightarrow o_2 \rightarrow i_4 \rightarrow o_4 \rightarrow i_2 \rightarrow o_1$	$k_1 \vee \bar{k}_3 \vee \bar{k}_2 \vee k_4$	Yes. Covered by 1
3	$o_1 \rightarrow i_3 \rightarrow o_3 \rightarrow i_2 \rightarrow o_1$	$k_1 \vee \bar{k}_4$	No
4	$o_1 \rightarrow i_3 \rightarrow o_4 \rightarrow i_1 \rightarrow o_1$	$k_1 \vee k_3$	No
5	$o_1 \rightarrow i_3 \rightarrow o_4 \rightarrow i_2 \rightarrow o_1$	$k_1 \vee k_4$	No
6	$o_1 \rightarrow i_4 \rightarrow o_4 \rightarrow i_1 \rightarrow o_1$	$k_2 \vee k_3$	No
7	$o_1 \rightarrow i_4 \rightarrow o_4 \rightarrow i_1 \rightarrow o_2 \rightarrow i_3 \rightarrow o_3 \rightarrow i_2 \rightarrow o_1$	$k_2 \vee k_3 \vee \bar{k}_1 \vee \bar{k}_4$	Yes. Covered by 6
8	$o_1 \rightarrow i_4 \rightarrow o_4 \rightarrow i_2 \rightarrow o_1$	$k_2 \vee k_4$	No
9	$o_2 \rightarrow i_3 \rightarrow o_3 \rightarrow i_1 \rightarrow o_2$	$\bar{k}_1 \vee \bar{k}_3$	No
10	$o_2 \rightarrow i_3 \rightarrow o_3 \rightarrow i_1 \rightarrow o_1 \rightarrow i_4 \rightarrow o_4 \rightarrow i_2 \rightarrow o_2$	$\bar{k}_1 \vee \bar{k}_3 \vee k_2 \vee k_4$	Yes. Covered by 9
11	$o_2 \rightarrow i_3 \rightarrow o_3 \rightarrow i_2 \rightarrow o_2$	$\bar{k}_1 \vee \bar{k}_4$	No
12	$o_2 \rightarrow i_3 \rightarrow o_3 \rightarrow i_2 \rightarrow o_1 \rightarrow i_4 \rightarrow o_4 \rightarrow i_1 \rightarrow o_2$	$\bar{k}_1 \vee \bar{k}_4 \vee k_2 \vee k_3$	Yes. Covered by 11
13	$o_2 \rightarrow i_3 \rightarrow o_4 \rightarrow i_1 \rightarrow o_2$	$\bar{k}_1 \vee k_3$	No
14	$o_2 \rightarrow i_4 \rightarrow o_4 \rightarrow i_1 \rightarrow o_2$	$\bar{k}_2 \vee k_3$	No
15	$o_2 \rightarrow i_4 \rightarrow o_4 \rightarrow i_1 \rightarrow o_1 \rightarrow i_3 \rightarrow o_3 \rightarrow i_3 \rightarrow o_2$	$\bar{k}_2 \vee k_3 \vee k_1 \vee \bar{k}_4$	Yes. Covered by 14
16	$o_2 \rightarrow i_4 \rightarrow o_4 \rightarrow i_2 \rightarrow o_1 \rightarrow i_3 \rightarrow o_3 \rightarrow i_1 \rightarrow o_2$	$\bar{k}_2 \vee k_4 \vee k_1 \vee \bar{k}_3$	Yes. Covered by 1

$k_1 \vee \bar{k}_3 \vee \bar{k}_2 \vee k_4$ is redundant since $k_1 \vee \bar{k}_3$ already covers it. Similarly, $k_2 \vee k_3 \vee \bar{k}_1 \vee \bar{k}_4$ is redundant since $k_2 \vee k_3$ covers it. Hence, the reduced cycle constraints are:

$$k_1 \vee \bar{k}_3$$

$$k_2 \vee k_3$$

For the example in Table 4.1, the last column indicates the *redundant constraints*. In order to remove all logically redundant cycle constraints, a simple method is used. First, noting that a redundant constraint always has more literals than the one covering it, the constraints are sorted in non-increasing order of their sizes in terms of the number of literals in each constraint. Let C^s denote the number of literals in the clause C. Each constraint C needs to be compared for possible redundancy only with constraints that have a smaller size than C^s.

Let the total number of key variables (i.e. total number of key bits of all the multiplexors combined) be W. Let the variables be indexed $1, 2, \ldots W$. In the constraint clauses, the key variables usually appear in the order in which the key-labeled edges in G' are traversed during *cycle generation*. Each clause is then rearranged such that the literals appear in increasing numerical order of their indices.

Algorithm 21: No-cycle constraints generation and reduction [25, 32]

Input: FEOL Circuit with Key Controlled MUX Network N
Output: Reduced No-Cycle Constraints S
1 Generate MUX graph G' from N;
2 Identify the set of all cycles in G' and generate the list of NC constraints S for the cycles
   in the form of Eq. 4.11;
3 Sort S in non-increasing order of the number of literals;
4 **for** $i = 1$ *to* $|S|$ **do**
5 | Sort the literals in clause $S[i]$ in increasing order of their indices;
6 **end**
7 **for** $i = 1$ *to* $|S|$ **do**
8 | **for** $j = i + 1$ *to* $|S|$ **do**
9 | | **if** $S[j]^s < S[i]^s$ **then**
10 | | | **if** $(S[i]^f \leq S[j]^f)$ *and* $(S[i]^l \geq S[j]^l)$ **then**
11 | | | | **if** *every literal in* $S[j]$ *also appears in* $S[i]$ **then**
12 | | | | | $S = S - S[i]$;
13 | | | | | **break**;   // break out of the innermost loop
14 | | | | **end**
15 | | | **end**
16 | | **end**
17 | **end**
18 **end**
19 **return** S;

For example, $(\bar{k}_3 \vee k_6 \vee \bar{k}_4 \vee k_2)$ will be rearranged as $(k_2 \vee \bar{k}_3 \vee \bar{k}_4 \vee k_6)$. After rearrangement, for a constraint clause C, let C^f denote the index of the first variable, C^l denote the index of the last variable, and $C^r = [C_f, C_l]$ denote the range of variable indices in C. That is, if $C = (k_2 \vee \bar{k}_3 \vee \bar{k}_4 \vee k_6)$, then $C^f = 2$, $C^l = 6$, and $C^r = [2, 6]$. For a clause C_2 to be redundant with respect to C_1, the range of C_2 should be the same as or subsume the range of C_1. Conversely, if this condition is not met, then there is no need to check for redundancy.

The algorithm for cycle optimization based on these observations is shown as Algorithm 21.

4.3.3 Attack Algorithm

An attack method based on the CycSAT attack proposed by Zhou et al. [23] which is in turn based on the SAT attack for encrypted logic circuits [18] is used to recover the correct key bits which reveal the BEOL connections. Algorithm 22 shows the attack procedure. N is the netlist of the circuit including the key-controlled MUX network shown in Fig. 4.6. $IC(I)$ is the IC or oracle which, given the inputs I, produces the values for outputs O. First, the NC constraints for N are generated and

---

Algorithm 22: SAT attack against SM [25] based on CycSAT [23]

Input: FEOL Circuit with Key Controlled MUX Network N, Oracle $IC(I)$
Output: MUX Key Values to Yield Correct BEOL Nets k

1 Generate and reduce the NC constraints set S for N using Algorithm 21;
2 Prepare CNF $C(I, K, O)$ for N;
3 $C(I, K, O) = C(I, K, O) \wedge S; ;$ // Add the no-cycle constraints.
4 Make two copies $C(I, K_A, O_A)$ and $C(I, K_B, O_B)$;
5 $F = C(I, K_A, O_A) \wedge C(I, K_B, O_B)$;
6 **while** $SAT(F \wedge (O_A \neq O_B))$ **do**
7   i = assignment to I determined by $SAT(F \wedge (O_A \neq O_B))$;
8      $o = IC(i)$;
9      $F = F \wedge C(i, K_A, o) \wedge C(i, K_B, o)$;
10 **end**
11 k = assignment to K_A determined by $SAT(F)$;
12 **return** k;

---

reduced as discussed before (line 1). Then, the CNF representation of the circuit is prepared (line 2). NC constraints are added to the circuit CNF clause set (line 3). The rest of the algorithm constitutes the standard SAT attack [18] discussed in Sect. 4.1.3. At the end, the algorithm returns a set of MUX control key values which when applied recover the original circuit or its logical equivalent. The connections made by the multiplexors yield the BEOL nets.

4.3.4 Discussion

Chen et al. [25] used the ISCAS-85 and ITC-99 benchmarks to evaluate the application of the SAT attack against SM. The benchmarks were bipartitioned using the PaToH tool [33] to identify the BEOL nets. Algorithm 22 correctly recovered the BEOL nets or their logical equivalents in all cases. Attack times ranged from a fraction of a second to 19 h. Number of SAT iterations (while loop in Algorithm 22) ranged from 12 to 216, number of clauses in the last iteration (the size of the largest SAT instance generated) ranged from 2000 to 7,000,000, and the number of Boolean variables in that last iteration ranged from 1000 to 25,000,000. For these benchmarks, the cutset sizes ranged from 17 to 114, key sizes ranged from 60 to 1480, number of NC constraints enumerated (before reduction) ranged from 0 to 19,560, and number NC constraints after reduction ranged from 0 to 8300 (reduction by up to 58%).

Success of the SAT based approach depends at least in part on the attacker's ability to identify the FEOL partitions so that the simplified MUX network shown in Fig. 4.5 can be used for recovery. If this is not possible, then the attacker is forced to use the general network shown in Fig. 4.4 which results in a much larger number of combinational cycles and, consequently, more NC constraints resulting in a larger SAT problem. Chen et al. noted that the current SAT solvers are unable to handle more than 1M NC constraints in under 48 h. In the absence of partitioning information, this limit would be reached for all but the 5 smallest benchmarks in the ISCAS-85 suite. Another way to simultaneously defeat the SAT based and the proximity based attacks is to avoid the min-cut partitioning to identify the BEOL nets and instead use a signal ranking function, such as the one discussed in Sect. 3.6, which would also lead to more cycles. When that method was used for BEOL net selection, only the 6 smallest benchmarks could be attacked by the SAT method in under 48 h.

The method proposed by Chen et al. can recover 100% of the BEOL nets without using any design constraints but requires an oracle for the underlying SAT attack. In addition, attack time and memory requirements for large SAT problems can prevent its successful application for practical problems.

4.4   SAT Attack Against Split Sequential Circuits

The reverse engineering method discussed in previous section is limited to combinational logic circuits. Chen et al. [32] proposed a method to apply the SAT attack against split manufactured *sequential circuits* by unrolling them into combinational logic circuits. The problem is formulated as follows:

A directed graph $G = (V, E)$ represents a sequential circuit. V can be partitioned into V_l and V_f representing the combinational logic gates and the flip-flops (FF) respectively. Similarly, E can be partitioned into E_l and E_f where E_l are all the connections between V_l nodes and E_f are the remaining edges involving at least one node in V_f. Hence, $G = (V, E) = (V_l \cup V_f, E_l \cup E_f)$. Nodes with no incoming (outgoing) edges represent the primary input (output) buffers. Let $H \subseteq E$ be the subset of nets selected to be assigned to the BEOL layers. Remaining circuit $(V, E - H)$ is assigned to the FEOL layers.

Due to the insecure foundry, the attacker already knows the logic and netlist of $(V, E - H)$. For a sequential circuit, let $I_{(x,t)}$ denote the input vector applied at cycle t, $0 \le t \le n - 1$ and let $O_{(x,t)}$, $1 \le t \le n$ be the output vectors produced by the circuit in response. The problem is to recover H correctly by using several (strategically selected) input–output sequences that the attacker is assumed to be able to obtain by having access to an oracle in the form of a packaged IC or a simulation model.

4.4.1   Attack Methodology

The FEOL sequential circuit is unrolled into an iterative combinational logic network over N successive time frames as shown in Fig. 4.7. In order to unroll

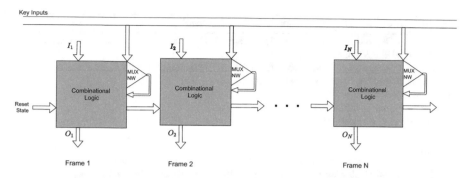

Fig. 4.7 Spatial unrolling of a FEOL sequential circuit

the sequential circuit, all FFs are removed and the inputs signal of the each FF in frame i is connected directly to the sink gates of that FF's output in frame $i + 1$. The input signals of FFs generated in frame N are ignored. The reset state is directly provided to the first frame. A *key-controlled interconnection network* similar to the one discussed in the previous section is inserted into each frame to recover the missing BEOL connections. Combinational cycles introduced by the interconnect network are handled by introducing and reducing the cycle constraints as discussed in the previous section. Since the BEOL connections remain the same in all frames, the key solutions will have to be the same in each frame. Hence, the same key inputs are used for all time frames shown in Fig. 4.7.

FEOL sequential circuits need to be unrolled over multiple cycles but the minimum unroll cycles to recover the correct keys is unknown. Though the attacker can unroll the sequential circuit for a large number of clock cycles, the attack time increases tremendously. In order to recover the set of correct BEOL connections as fast as possible, the minimum number of unrolling frames is preferred. Algorithm 23 shows the steps in the attack methodology. A key-controlled interconnect network is inserted in the FEOL circuit to determine the potential BEOL connections as discussed in Sect. 4.3. Structural cycles introduced by FFs are removed by disconnecting the FF input signals and combinational cycle constraints C are identified using this circuit. Then, the cycle optimization method removes all logically redundant cycle constraints and the simplified cycle constraints are generated for the next step. The combinational circuit G_N is generated by unrolling the sequential circuit over N clock cycles where $N \geq 2$. The cycle constraints enabled SAT attack discussed in Sect. 4.3 is used to recover key values K_N. If the circuit with BEOL connections implied by key K_N is determined to be equivalent to the original circuit, the attack has succeeded. Otherwise, the unroll frame size N will be increased and the process is repeated.

In the attack method mentioned above, the attacker cannot perform sequential equivalence checking with the original circuit to validate the recovered keys. However, the attacker can validate the recovered circuit "at-speed" with the help of an FPGA by implementing the recovered circuit in the FPGA, applying the same

---
Algorithm 23: SAT attack against split sequential circuits [32]

Input: FEOL Netlist $G = (V, E - H)$, Packaged IC (*eval*), Maximum Frames N_{max}
Output: BEOL Connections H
1 Insert a key-controlled multiplexor network, with unknown key inputs K, between the
    FEOL output terminals and input terminals.
2 Remove the flip-flops from V.
3 Generate cycle constraints, C.
4 Simplify the cycle constraints C.
5 N = 2;
6 **repeat**
7 | Unroll the circuit into N frames to obtain combinational circuit G_N;
8 | Use the SAT attack on G_N including constraints C to obtain key values K_N;
9 | Construct FEOL circuit G with BEOL connectivity H extracted from K_N;
10 | $N = N + 1$;
11 **until** *G is correctly validated or $N > N_{max}$*;
12 **return** H;

---

input sequences to both the emulated circuit and the oracle, and comparing their
primary outputs over sufficiently large (or desired) sequences of input vectors.

4.4.2 Discussion

Chen et al. [32] evaluated the performance of Algorithm 23 using sequential circuits
selected from the ISCAS-89 and ITC-99 suites. The netlists were bipartitioned using
[33] to identify the BEOL nets. Then, these nets were removed and the resulting
FEOL circuit was used as input to Algorithm 23. Combinational equivalence
checking using a single frame of combinational logic, while treating flip-flops as
primary inputs/outputs, is used to verify the correctness of the recovered circuit. This
is a stronger verification than necessary since the circuit reset state is not considered.

The selected benchmarks contained up to 20,027 gates and 1728 flip-flops. The
number of BEOL nets ranged from 13 to 90 and the number of key bits for the MUX
network ranged from 48 to 540. Cycle optimization step reduced the number of NC
constraints by up to 99.85% (from 492,650 to 722). Algorithm 23 recovered all of
the BEOL connections or their functional equivalents correctly requiring up to 9
unrolls and up to an hour execution time.

Since this attack requires unrolling a sequential circuit potentially over a large
number of cycles, its applicability is limited to relatively small circuits in practice.

4.5 SAT Attacks Including Proximity Information

SAT attacks against SM as discussed in the previous sections are limited to small circuits since the size of the key network is dependent directly on the number of dangling source and sink pins in the FEOL but not on their proximities or any other information extracted from the layout. Chen et al. [34] proposed a simple method to exploit *proximity information* to reduce the size of the MUX network and significantly speed up the SAT attack.

4.5.1 Proximity Information

The proximity attack is a sequential connection recovery attack which recovers the BEOL connections one by one. The priority function used in the proximity attacks is the shortest distance criterion between BEOL input/output terminals [27]. This criterion is motivated by the assumption that placement tools try to keep the connected terminals as closely as possible in order to minimize the overall wire length. However, over-dependence on the shortest wire length criterion can mislead the recovery algorithm since in many circumstances the correct connection for a target cell is not necessarily the nearest terminal in the candidate list. In the presence of conflicts and routing congestion, physical design algorithms use many heuristics to break ties. In addition, the cost function is the estimated *total* wire length rather than the length of each net. Wire length estimation is not necessarily based on rectilinear distances between source and sink nodes, especially for multi-terminal nets [35]. Fixed I/O pin positions may preclude wire length minimal placements. Finally, performance-driven layout synthesis tools give priority to bring together critical signals/gates in critical paths while allowing nodes in non-critical paths to be spread apart. These are the nets that are likely to be assigned to the upper BEOL metal layers.

Even when the BEOL signals are given high priority, it is not possible to place all of the corresponding nodes in close proximity. Consider a gate v_i with two BEOL input signals from gates v_x and v_y. Let l_x and l_y be the estimated wire lengths of the respective wires. If $l_x < l_y$ and the attack is trying to recover a connection for v_y first, then an incorrect decision may be made from which a sequential attack may not recover. Referring to Fig. 4.8, if sequential proximity attack chooses to recover a net from o_2 before considering the other output terminals, a connection between o_2 and i_3 will be inferred since i_3 is the closest input terminal to o_2. Once an incorrect decision is made, more incorrect connections follow for the rest of the attack as a sequence of recovery decisions are made. Hence, the connection between o_2 and i_2 may never be considered since o_2 is already connected with i_3 or such a consideration would require significant *backtracking*.

In spite of these caveats, the proximity information can be quite useful when multiple nets are *simultaneously* considered. Let n be the number of missing BEOL

Algorithm 24: Relative distance computation

Input: FEOL input terminals I, FEOL output terminals O
Output: Relative distances r

1 **foreach** $i \in I$ **do**
2 $\quad W_i$ = output nodes $o \in O$ arranged in non-decreasing order of $w(i, o)$;
3 $\quad r(i, W_i[1]) = 1$;
4 $\quad$ **for** $k = 2$ *to* $|O|$ **do**
5 $\quad\quad$ **if** $w(i, W_i[k]) = w(i, W_i[k-1])$ **then**
6 $\quad\quad\quad | \quad r(i, W_i[k]) = r(i, W_i[k-1])$;
7 $\quad\quad$ **else**
8 $\quad\quad\quad | \quad r(i, W_i[k]) = r(i, W_i[k-1]) + 1$;
9 $\quad\quad$ **end**
10 $\quad$ **end**
11 **end**
12 **return** r;

---

signals. For simplicity, we assume that these all are two terminal nets. Hence, the number of BEOL output/input nodes is n. For each BEOL input node i and output node o, the distance between i and o can be computed using the placement information available to the attacker due to the untrusted foundry where FEOL layers are fabricated. Let $w(i, o)$ denote this distance. Then, for each BEOL input node i, a sorted list W_i of BEOL output nodes in ascending order of their distances from i is prepared. For each i, $1 \leq i \leq n$, the *relative distance* $r(i, o)$ from i to o is computed as shown in Algorithm 24.

For each input node i, the relative distance relation induces an equivalence class on the output nodes. Clearly, with respect to any input node, relative distance numbers for the output nodes will be in the range 1 to n. For a given input node i and relative distance d, we define the *Reach Set*, $R(i, d)$ as the set of all output nodes o with relative distance $r(i, o) \leq d$.

For example, for the terminals in Fig. 4.9, we assume the correct connection for input terminal i_8 is o_7 while the nearest available output terminal is o_4. This is quite common since not all paths can be optimized during layout synthesis. Suppose, the wire lengths between i_8 and the various output nodes are as follows: $w(i_8, o_4) < w(i_8, o_7) = w(i_8, o_6) = w(i_8, o_8) < w(i_8, o_2) = w(i_8, o_9) <$

Fig. 4.8 Placement proximity example

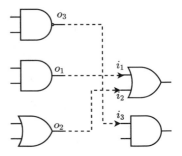

Fig. 4.9 Proximity
measurement

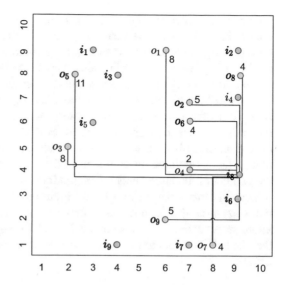

$w(i_8, o_1) = w(i_8, o_3) < w(i_8, o_5)$. We can calculate the relative distance of all available output terminals from i_8 as follows:

- $r(i_8, o_4) = 1$
- $r(i_8, o_7) = r(i_8, o_6) = r(i_8, o_8) = 2$
- $r(i_8, o_2) = r(i_8, o_9) = 3$
- $r(i_8, o_1) = r(i_8, o_3) = 4$
- $r(i_8, o_5) = 5$

4.5.2 Reduction of Key-Controlled Network

SAT attack for SM begins by modeling the missing BEOL signals using a key-controlled interconnection network as shown in Fig. 4.4 and determines the correct key values using a SAT solver. This explores $n!$ connection combinations when the total number of nets in BEOL is n and all nets are 2-terminal connections. If the attacker can determine that the BEOL signal removal induces a bipartition on FEOL circuit and the partition to which each BEOL terminal belongs, as assumed in [27], then the number of connection combinations can be reduced to $2(\frac{n}{2})!$.

By making use of the Reach information previously defined, the attacker can significantly reduce the network size. Suppose the attacker makes a guess that the correct output node driving any input node is within a relative distance d. Then, the interconnect model can exclude all output nodes which are farther away than d from being considered by any key combination. This significantly reduces the size of the network and the number of keys for the SAT attack. In practice, it is impossible for the attacker to correctly guess the exact relative distance to include all the correct connections. However, the attacker is able to verify whether the recovered circuit is correct using large number of simulations/emulations. (Researchers can verify this by formal equivalence checking since they know the correct logic circuits.) Hence, if the correct connections are not included in the interconnect network, then the attacker will fail to generate a candidate circuit or will generate an incorrect circuit. In this case, the attacker can increase the bound on the relative distance, construct a larger interconnect network, and repeat the SAT attack. In the worst case, the bound on the relative distance will have to reach the maximum value for all terminals, that is, there would be no reduction in the network size or number of keys. Thus, the number of the interconnect combinations is 1 in the best case (corresponding to relative distance bound 1 and for all input terminals i, $R(i, 1) = 1$) and $2(\frac{n}{2}!)$ in the worst case. But in most cases, experimental results suggest that the number of interconnect combinations is bounded by a^n where a is a small constant, indicating that the BEOL output terminal driving each input terminal is within a small relative distance.

4.5.3 Attack Algorithm for Combinational Circuits

The proximity information is used to reduce the key network size for the SAT attack. The detailed attack algorithm is shown as Algorithm 25. In step 1, the wire length between each output terminal with each input terminal is calculated, and then relative distances for all output terminals from each input terminal are generated. Using this information, Reach Sets for each input terminal and each relative distance are obtained. In steps 4–5, candidate output terminals for each input terminal with relative distance of d or less are selected and a key-controlled interconnect network is generated and inserted to the FEOL circuit with these selected candidate terminals. From 6–8, all combinational cycles in the circuit generated in the previous step are located and cycle constraints are generated and optimized. With the circuit logic and the cycle constraints, the SAT attack [18] is used to recover the correct connections. In step 9, if the keys are recovered correctly, then the correct BEOL connections are returned. Otherwise, relative distance is increased by 1 and steps 4–9 are repeated until the relative distance reaches its maximum value. BEOL recovery is guaranteed when d is maximum since all possible output terminals are included for each input terminal. However, the algorithm usually terminates for a much smaller value of d.

4.5.4 Attack Algorithm for Sequential Circuits

The detailed attack flow for sequential circuits is shown in Algorithm 26. In steps 1–7, the attack process is quite similar to the combinational circuit attack process,

---

Algorithm 25: Improved SAT attack algorithm against SM [34]

Input: FEOL layers $(V, E - H)$, Packaged IC or equivalent oracle $eval()$
Output: Correct BEOL Key K_c
1   For all input nodes i and relative distances d, determine Reach Sets $R(i, d)$;
2   $d = 0$;
3   **repeat**
4 | $d = d+1$;
5 | Construct a combinational circuit with key-controlled interconnection network N_d
    | with keys K_d to allow all possible connections between each BEOL input node i and
    | any BEOL output node o such that $o \in R(i, d)$;
6 | Construct all combinational cycles C_d in N_d;
7 | Eliminate redundant cycles from C_d;
8 | Generate a SAT attack problem S_d to find K_d values in N_d with the constraints C_d
    | such that K_d are compatible with eval();
9   **until** *(S_d is successfully solved by the SAT attack)*;
10 Report the BEOL signals H corresponding to the K_d values returned by the successful
    SAT attack;

Algorithm 26: Improved sequential SAT attack algorithm against SM [34]

Input: FEOL layers $(V, E - H)$, Packaged IC or equivalent oracle $eval()$
Output: Correct BEOL Key K_d
1   For all input nodes i and relative distances d, determine Reach Sets $R(i, d)$;
2   $d = 0$;
3   **L1:**
4   $d = d + 1$;
5   Construct a sequential circuit with key-controlled interconnection network N_d with keys
    K_d to allow all possible connections between each BEOL input node i and any BEOL
    output node o such that $o \in R(i, d)$;
6   Construct all combinational cycles C_d in N_d;
7   Eliminate redundant cycles from C_d;
8   $u = 1$;
9   **L2:**
10 $u = u + 1$;
11 Unroll the sequential circuit N_d over u clock cycles and form a combinational equivalent
    circuit $N_d(u)$ with keys K_d applied to all the slices;
12 Generate a SAT attack problem S_d to find K_d values in $N_d(u)$ with the constraints C_d such
    that K_d are compatible with eval();
13 **if** *(SAT attack on S_d is not solvable)* **then**
14 | **Goto L1**;
15 **else**
16 | Verify the solution for K_d obtained by the SAT attack;
17 | **if** K_d *is incorrect* **then**
18 | | **Goto L2**;
19 | **else**
20 | | Report the BEOL signals H corresponding to the K_d values returned by the
    | | successful SAT attack;
21 | **end**
22 **end**

---

except that the signals involving the FFs are removed before cycle identification. In steps 10–12, the sequential circuit is unrolled over u clock cycles and a SAT problem is generated from the unrolled circuit and the cycle constraints resulting from steps 1–7. If the SAT attack does not succeed (step 13), it means that the correct connection combination is not included in the sized interconnect network. Hence, the proximity distance is increased and process is repeated from L1 (step 14). If the SAT attack is successful, then the key K_d generated should be verified. While the researchers can use formal verification, the attacker needs to use simulation/emulation based verification. If the key proves to be incorrect, then the *unroll size* (u) is increased and the process is repeated from L2 (step 18). If the key value is found to be acceptable, then it is reported (step 20). Although in the worst case this process may continue until the farthest proximity information (d_{max}) is included in the network formation and the unroll size u reaches the *deepest encrypted state*, in practice it will converge for relatively small values of d and u.

4.5.5 Discussion

Chen et al. [34] used 14 combinational and 10 sequential circuits from the ISCAS-85, ISCAS-89, and ITC-99 suites to evaluate the hybrid attacks. PaToH partitioning engine was used to identify the BEOL [33] to be removed and recovered. A force-directed placement algorithm [35] was used for placement. All the missing BEOL nets were correctly recovered by the attacks. For combinational circuits, the attack time reduced by a factor of up to 46.7× when proximity information was used with an average reduction of about 9×. Relative distance d at which the keys for the correct BEOL signals were found ranged from 2 to 10. For sequential circuits, the attack time reduced by a factor of up to 79.5× with the average reduction being about 14×. Successful d value ranged from 2 to 8 and u value ranged from 2 to 10.

Spreading the pins on a BEOL net farther apart in the FEOL layout forces an increase in the d value. This in turn increases the multiplexor size which in turn increases the number of NC constraints thereby increasing the size of the SAT problem. This observation can be used as a defense method. The pins of the missing BEOL nets which are not on the critical paths can be spread father apart to increase the effective d value. It is also possible to increase the number of BEOL nets, at the expense of BEOL fabrication cost, to increase the MUX key bits.

Chen et al.'s work demonstrated that hybrid attack methods combining design constraints with satisfiability based methods hold much promise when an oracle is available. Besides proximity, additional design constraints such as drive strength matching, direction of dangling wires, etc. can be considered to further reduce the key-controlled MUX network size.

4.6 SMT Based Reduction of Key Network Complexity

Reduction of the key-controlled MUX network size is important to reduce the attack time and increase the size of the designs that can be handled. Wang et al. [29] proposed a SAT attack against bipartitioned 2.5D SM designs based on introducing a MUX or DEMUX based key network similar to the one used by Chen et al. [25]. In addition, to reduce the size of the SAT problem, Wang et al. proposed additional hints to "group" the MUX outputs such that the total number of keys is reduced. The grouping method uses a *Satisfiability Modulo Theories* (SMT) [36] constraint formulation as described in this section. Use of SMT allows higher level constraints (e.g. m of the n bits in a vector should be 1's) to be easily added to the formula while making use of an underlying SAT engine to solve the constraints.

4.6.1 Hard Grouping Hints

Assume that the logic circuit is bipartitioned to identify the BEOL nets. Let N_1 and N_2 be the two partitions of the original combinational logic netlist N. Let the cutset size, that is, the number of hidden nets from N_1 to N_2, be n. For simplicity, assume that each hidden net is fanout-free, that is, each net has one output terminal from N_1 and one input terminal in N_2. Let these outputs of N_1 be represented by Boolean vector U and the inputs of N_2 be V, where $|U| = |V| = n$. Let vectors X and Y represent the primary inputs and primary outputs of N. The goal of the method is to find a *hard grouping* defined as a mapping $G : U \rightarrow \mathcal{P}(V)$ such that for all $u \in U$, $G(u) \subseteq V$ is guaranteed to include the actual input in V to which u is connected.

The algorithm searches through the space of all binary assignments to U in which h of the n bits are preset to 1 or 0. These preset bits are denoted as *hot bits* and h is explored iteratively. Let $t \in \{0, 1\}$ denote the preset value of the hot bits. Let $S(X, U, V, Y)$ denote the SMT formula representing the netlist consisting of N_1 and N_2 available to the attacker via the untrusted foundry. Similar to the methods discussed in the previous sections, the algorithm uses an oracle, denoted by $eval(X)$, to determine the Y values given the X values. HW denotes the Hamming Weight function which is defined as the number of 1's in a given binary vector.

Consider the example in Fig. 4.10 in which $n = 5$. Suppose, using the netlist and eval(), a mutually compatible set of vectors for X, Y, and U are found. Let $U = \langle 11000 \rangle$. Suppose $V = \langle 00110 \rangle$ is the assignment to V which is compatible with the same Y. This implies that, as shown by the dashed lines in Fig. 4.10, $G(u_1) = G(u_2) = \{v_3, v_4\}$ and $G(u_3) = G(u_4) = G(u_5) = \{v_1, v_2, v_5\}$. If there are multiple assignments to V which are compatible with the same (X, U, Y) vectors, then the union of groupings yields the final grouping for this U. Likewise, for each U vector, a grouping is determined and the intersection of all the groupings is the final grouping.

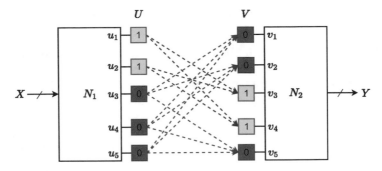

Fig. 4.10 Hard grouping example (based on [29])

4.6.2 Algorithm for Hard Grouping

Given S, h, and t, Algorithm 27 shows the hard grouping method. The algorithm formulates an SMT constraint to find a mutually compatible set of assignments to X, Y, U and V such that exactly h bits in U and V are of value t and the assignments to X and Y are validated by the actual observations. If such an assignment is found by SMT satisfiability, let U_i and X_i denote the assignments to U and X respectively (lines 10–11). From this, iteratively, all assignments V_j to V which are compatible with X_i and U_i are generated (lines 16–28). For each V_j generated, subgroups of bit positions in V which have the same values are extracted and recorded as potential maps to the bit positions in U that have the same values in the vector U_i (lines 18–24). The next compatible assignment to V is constrained such that it should not produce the same solution as in the previous iteration and, in addition, it should further refine the grouping (lines 25–26). Such a solution is denoted as a *distinguishing V vector* and will be explained through an example. This entire process continues as long as new qualifying assignments to U can be found (lines 9–31). At the end, the intersection of 0-propagating and 1-propagating groups yields the final mapping for each position in U.

The idea of a distinguishing V vector can be explained through the example. For a given U vector, a new V vector is useful only if it reveals new grouping information. Suppose $U = \langle 01100 \rangle$ and two compatible V vectors are $V = \langle 00110 \rangle$ and $V = \langle 01010 \rangle$. These values imply that the groupings for the bit positions assuming a value of 1 in U be $G(u_2) = G(u_3) = \{v_2, v_3, v_4\}$. If the next compatible V value is $V = \langle 01100 \rangle$, these two groupings remain unchanged. However, if the next compatible vector is $V = \langle 10100 \rangle$, then this reveals new grouping information and the groupings are changed to $G(u_2) = G(u_3) = \{v_1, v_2, v_3, v_4\}$. A V vector assignment that reveals new information is said to be a *distinguishing* vector. Line 25 in the algorithm adds a constraint that the next vector be a distinguishing vector for either bit positions in U assuming a 1 value or those assuming a 0 value or both. Rather than blindly enumerating all compatible vectors, addition of this constraint speeds up the search process significantly.

Algorithm 27: Hard signal grouping based on SMT solving [29]

---

1 find_group_with_hot_bits *(S, h, t)*

    Input: SMT Formula $S(X, U, V, Y)$, Number of Hot Bits h, Type of Hot Bits: t
    Output: Groups G

2 $i = 1$;
3   **if** $t = 1$ **then**
4 | $H = $ SMT formula for$((HW(U) = h) \wedge (HW(V) = h))$;
5   **else**
6 | $H = $ SMT formula for$((HW(U) = n - h) \wedge (HW(V) = n - h))$;
7   **end**
8 $F = S(X, U, V, Y) \wedge H \wedge (Y = eval(X))$;
9   **while** *satisfiable(F)* **do**
10 $U_i = $ SMT assignment for U in the solution of F;
11 $X_i = $ SMT assignment for X in the solution of F;
12 $F_{new} = F \wedge (U = U_i) \wedge (X = X_i)$;
13 $j = 1$;
14 $G_1(u) = \emptyset, \forall u \in U$;
15 $G_0(u) = \emptyset, \forall u \in U$;
16   **while** *satisfiable(F_{new})* **do**
17 $V_j = $ SMT assignment for V in the solution of F_{new};
18   **foreach** $u \in U$ **do**
19   **if** $U_i[u] = 1$ **then**
20 | $G_1(u) = G_1(u) \cup \{v \in V | V_j(v) = 1\}$;
21   **else**
22 $G_0(u) = G_0(u) \cup \{v \in V | V_j(v) = 0\}$;
23   **end**
24   **end**
25 $H = $ SMT formula for"V should be a new distinguishing vector";
26 $F_{new} = F_{new} \wedge H$;
27 $j = j + 1$;
28   **end**
29 $F = F \wedge (U \neq U_i)$;
30 $i = i + 1$;
31   **end**
32   **foreach** $u \in U$ **do**
33 | $G(u) = G_1(u) \cap G_0(u)$;
34   **end**
35   **return** G;
36   ——————————————————————————————
37 find_group *(S)*

    Input: SMT Formula $S(X, U, V, Y)$, h_{max}, h_{min}
    Output: Groups G

38 $h = h_{max}$;
39 $G(u) = \emptyset, \forall u \in U$;
40   **while** $h > h_{min}$ **do**
41 $G_{1i} = $ find_group_with_hot_bits$(S, h, 1)$;
42 $G_{0i} = $ find_group_with_hot_bits$(S, h, 0)$;
43   **foreach** $u \in U$ **do**
44 | $G(u) = G(u) \cap G_{1i}(u) \cap G_{0i}(u)$;
45   **end**
46   **if** $\forall u, |G(u)| \leq h_{min}$ **then**
47 | **break**;
48   **else**
49 | $h = h - 1$;
50   **end**
51   **end**
52   **return** G;

After the groupings are obtained, a reduced MUX network, with one key-controlled MUX for each group, is generated and used in the SAT attack. Different h and t values yield different grouping results. Smaller h values reduce the final SAT attack time but increase the grouping time. In addition, very small h values may not yield successful groupings at all. Procedure find_groups shown in Algorithm 27 explores the space of these values within a given range $[h_{min}, h_{max}]$. It starts with $h = h_{max}$ and explores the groupings with both t values. The intersections of the groupings in each bit position yield the grouping results for that iteration. If each group size is less than h_{min}, then the problem size is considered sufficiently reduced for the SAT attack. Experimentally, 50% of n and 20% of n are found to be reasonable upper and lower exploration bounds on h.

BEOL nets with fanout, that is nets which have a source in N_1 and multiple sinks in N_2 (in which case, $|V| \geq |U|$) can be easily incorporated in Algorithm 27. To do so, lines 4 and 6 should be modified to constrain V such that $(HW(V) \geq HW(U)) \wedge (((|V| - HW(V)) \geq (|U| - HW(U))))$.

4.6.3 Soft Grouping Hints

Wang et al. [29] have also suggested a *soft grouping strategy* to simplify the MUX network based on hints from the physical design of the FEOL part available to the attacker. Such hints may be from the physical proximity, timing slacks, and other information discussed in Chap. 2. Soft grouping approach leads to attack strategies similar to the hybrid attack proposed by Chen et al. [34] and discussed in Sect. 4.5.

4.6.4 Discussion

Wang et al. [29] evaluated the impact of SMT based grouping using some of the ISCAS-85 and MCNC benchmarks. For fanout-free circuits with cut sizes in the range 108–346, hot bits were in the range 50–163. Compared to the SAT attack without groupings, the SAT attack followed by groupings resulted in a net speed-up ranging from 13.8× to upwards of 1440×. When fanout is considered, only one benchmark could be finished in 24 h without grouping hints. In this case, the attack speed-up with grouping hints versus without ranged from 22.1× to at least 212.3×.

Wang et al. showed that grouping the signals based on logical and physical observations reduces the size of the MUX networks and, consequently, the size of the SAT problem.

Table 4.2 Summary of satisfiability based attacks

Sl.	Attack	Year	Type	Benchmarks	Metrics
1	Satisfiability based layout recognition	2013	TI	ISCAS-85, ITC-99	k-security
2	SAT attack based reverse engineering	2019	RE	ISCAS-85, ITC-99	AC
3	SAT attack against split sequential circuits	2019	RE	ISCAS-89, ITC-99	AC
4	SAT attacks including proximity information	2019	RE	ISCAS-85, ISCAS-89, ITC-99	AC
5	SMT based reduction of key network complexity	2020	RE	ISCAS-85, MCNC	AC

RE reverse engineering attack, *TI* trojan insertion attack

4.7 Summary

In this chapter, we have discussed several SAT based attacks in which the attack problem is modeled as a Boolean satisfiability problem. Table 4.2 shows a summary of these attacks. Although they are effective in ensuring 100% attack success, they are limited by the attack time and the need to have access to an oracle. Attack time can be reduced significantly by mounting hybrid attacks in which the size of the SAT problem is reduced by considering design constraints. This approach holds much promise for future research.

References

1. A. Biere, M. Heule, H. van Maaren, T. Walsh, *Handbook of Satisfiability*, vol. 3 (IOS Press, Amsterdam, 2009)
2. D. Knuth, *The Art of Computer Programming. Volume 4, Fascicle 6: Satisfiability*, vol. 2 (Addison-Wesley, Reading, 2015)
3. U. Schöning, J. Torán, *The Satisfiability Problem: Algorithms and Analyses* (Lehmanns Media, 2013)
4. V.W. Marek, *Introduction to Mathematics of Satisfiability* (Chapman & Hall/CRC, London, 2009)
5. M.R. Garey, D.S. Johnson, *Computers and Intractability: A Guide to the Theory of NP-Completeness* (W. H. Freeman, New York, 1979)
6. T. Balyo, M.J. Froleyks, N. Heule, M. Iser, M. Järvisalo, M. Suda, Proceedings of SAT Competition 2020: Solver and Benchmark Descriptions. University of Helsinki, Department of Computer Science, 2020
7. M. Davis, G. Logemann, D. Loveland, A machine program for theorem-proving. Commun. ACM **5**(7), 394–397 (1962)
8. J.P. Marques-Silva, K.A. Sakallah, GRASP: a search algorithm for propositional satisfiability. IEEE Trans. Comput. **48**(5), 506–521 (1999)

9. F.V. Andrade, L.M. Silva, A.O. Fernandes, Improving SAT-Based combinational equivalence checking through circuit preprocessing, in *26th IEEE International Conference on Computer Design 2008, ICCD* (2008), pp. 40–45

10. E. Clarke, A. Biere, R. Raimi, Y. Zhu, Bounded model checking using satisfiability solving. Formal Methods Syst. Des. **19**(1), 7–34 (2001)

11. M. Yasin, J. Rajendran, O. Sinanoglu, *Trustworthy Hardware Design: Combinational Logic Locking Techniques*, ser. Analog Circuits and Signal Processing (Springer, Berlin, 2020)

12. J.A. Roy, F. Koushanfar, I.L. Markov, EPIC: Ending piracy of integrated circuits, in *Proceedings -Design, Automation and Test in Europe, DATE* (2008), pp. 30–38

13. A. Baumgarten, A. Tyagi, J. Zambreno, Preventing IC piracy using reconfigurable logic barriers. IEEE Des. Test Comput. **25**(1), 66–75 (2010)

14. J. Rajendran, Y. Pino, O. Sinanoglu, R. Karri, Security analysis of logic obfuscation, in *Proceedings—Design Automation Conference* (2012), pp. 83–89

15. S. Dupuis, P.S. Ba, G. Di Natale, M.L. Flottes, B. Rouzeyre, A novel hardware logic encryption technique for thwarting illegal overproduction and Hardware Trojans, in *Proceedings of the 2014 IEEE 20th International On-Line Testing Symposium, IOLTS 2014* (2014), pp. 49–54

16. J. Rajendran, H. Zhang, C. Zhang, G.S. Rose, Y. Pino, O. Sinanoglu, R. Karri, Fault analysis-based logic encryption. IEEE Trans. Comput. **64**(2), 410–424 (2015)

17. M. Yasin, J.J. Rajendran, O. Sinanoglu, R. Karri, On improving the security of logic locking. IEEE Trans. Comput. Aided Des. Integr. Circuits Syst. **35**(9), 1411–1424 (2016)

18. P. Subramanyan, S. Ray, and S. Malik, "Evaluating the security of logic encryption algorithms, in *Proceedings of the IEEE International Symposium on Hardware-Oriented Security and Trust, HOST* (2015), pp. 137–143

19. M. Yasin, B. Mazumdar, J.J. Rajendran, O. Sinanoglu, SARLock: SAT attack resistant logic locking, in *Proceedings of the IEEE International Symposium on Hardware Oriented Security and Trust* (2016), pp. 235–241

20. Y. Xie, A. Srivastava, Anti-SAT: mitigating sat attack on logic locking. IEEE Trans. Comput. Aided Des. Integr. Circuits Syst. **38**(2), 199–207 (2019)

21. M. Yasin, A. Sengupta, M.T. Nabeel, M. Ashraf, J. Rajendran, O. Sinanoglu, Provably-Secure logic locking: From theory to practice, in *Proceedings of the ACM Conference on Computer and Communications Security* (2017), pp. 1601–1618

22. K. Shamsi, M. Li, T. Meade, Z. Zhao, D. Z. Pan, Y. Jin, Cyclic obfuscation for creating SAT-unresolvable circuits, in *Proceedings of the ACM Great Lakes Symposium on VLSI*, vol. Part F127756, 2017, pp. 173–178

23. H. Zhou, R. Jiang, S. Kong, CycSAT: SAT-based attack on cyclic logic encryptions, *IEEE/ACM International Conference on Computer-Aided Design, Digest of Technical Papers, ICCAD*, vol. 2017-Novem, 2017, pp. 49–56

24. F. Imeson, A. Emtenan, S. Garg, M.V. Tripunitara, Securing computer hardware using 3D integrated circuit (IC) technology and split manufacturing for obfuscation, in *Proceedings of the 22nd USENIX Security Symposium* (USENIX Association, Berkeley, 2013), pp. 495–510

25. S. Chen, R. Vemuri, On the Effectiveness of the Satisfiability Attack on Split Manufactured Circuits," *IEEE/IFIP International Conference on VLSI and System-on-Chip, VLSI-SoC*, vol. 2018-Octob, 2019, pp. 83–88

26. J.F.C. Kingman, V.E. Benes, *Mathematical Theory of Connecting Networks and Telephone Traffic*, vol. 2 (Academic Press, London, 1966)

27. J. Rajendran, O. Sinanoglu, R. Karri, Is Split Manufacturing Secure? in *Proceedings—Design, Automation and Test in Europe, DATE* (2013), pp. 1259–1264

28. Y. Xie, C. Bao, A. Srivastava, Security-aware 2.5D integrated circuit design flow against hardware IP piracy. Computer **50**(5), 62–71 (2017)

29. W.C. Wang, Y. Wu, P. Gupta, Reverse engineering for 2.5-D split manufactured ICs. IEEE Trans. Comput. Aided Des. Integr. Circuits Syst. **39**(10), 3128–3133 (2020)

30. S. Roshanisefat, H.M. Kamali, A. Sasan, SRCLock: SAT-Resistant cyclic logic locking for protecting the hardware, in *Proceedings of the ACM Great Lakes Symposium on VLSI* (2018), pp. 153–158

31. D.B. Johnson, Finding all the elementary circuits of a directed graph. SIAM J. Comput. **4**(1), 77–84 (1975)
32. S. Chen, R. Vemuri, Reverse engineering of split manufactured sequential circuits using satisfiability checking, in *Proceedings—2018 IEEE 36th International Conference on Computer Design, ICCD 2018* (2019), pp. 530–536
33. Ü. Çatalyürek, C. Aykanat, PaToH (partitioning tool for hypergraphs), in *Encyclopedia of Parallel Computing* (Springer, Berlin, 2011), pp. 1479–1487
34. S. Chen, R. Vemuri, Exploiting proximity information in a satisfiability based attack against split manufactured circuits, *Proceedings of the 2019 IEEE International Symposium on Hardware Oriented Security and Trust, HOST 2019* (2019), pp. 171–180
35. N.A. Sherwani, *Algorithms for VLSI Physical Design Automation* (Springer, Berlin, 2012)
36. L. De Moura, N. Bjørner, Z3: An efficient SMT Solver, in *Lecture Notes in Computer Science (including subseries Lecture Notes in Artificial Intelligence and Lecture Notes in Bioinformatics)*, vol. 4963 LNCS (Springer, Berlin, 2008), pp. 337–340

Chapter 5
Defenses Against Satisfiability Based Attacks

Abstract Satisfiability based attacks discussed in the previous chapter formulate the attack as a satisfiability problem and use SAT solvers to find a netlist to layout mapping or to recover the missing BEOL nets. In this chapter, we discuss several defense methods to thwart the SAT attacks. In general, these methods attempt to increase the computational cost of the SAT attack to the point where it is no longer feasible. Several of these defense methods can defend the IC design against both SAT attacks and design constraint attacks. Some of these methods effectively combine other defense techniques such as logic locking and layout camouflaging with SM to improve the overall security against multiple attack models. We discuss the following defense methods: greedy wire lifting based on satisfiability, simultaneous wire lifting and cell insertion, combined layout camouflaging and SM for 2D and 3D designs, combined logic locking and SM, and combined SM and DFT (Design for Testability).

In the previous chapters, we have discussed design constraint based attacks and satisfiability based attacks against split circuits. These attacks aim to insert hardware trojans into the FEOL layouts or recover the missing BEOL nets. We have also discussed several defense methods to design ICs to thwart attacks that exploit design constraints. In this chapter, we will discuss defense methods to thwart satisfiability based trojan insertion attacks. In addition, several of these methods can also mitigate design constraint based reverse engineering attacks.

5.1 SAT Based Greedy Wire Lifting

In a trojan insertion or layout recognition attack, the attacker is interested in locating a node in the layout that matches specific circuit structures with the intent to insert a trojan at that node. In Sect. 4.2, we have discussed a satisfiability based method for finding in a graph a subgraph that is isomorphic to another given graph and explained how this method, due to Imeson et al. [1], can be used for a layout

© The Author(s), under exclusive license to Springer Nature Switzerland AG 2021
R. Vemuri, S. Chen, *Split Manufacturing of Integrated Circuits for Hardware Security and Trust*, https://doi.org/10.1007/978-3-030-73445-9_5

recognition attack. In this section, we will discuss a wire lifting method, due to
Imeson et al., which aims to ensure a desired level of k-security.

5.1.1 Objective

Imeson et al. [1] proposed the concept of k-security (discussed in Sect. 1.12) in the
context of layout recognition attacks. k-security can be formalized as follows:

Let $G = (V_g, E_g)$ be the netlist graph and let $H = (V_h, E_h)$ be a subgraph where
$V_h = V_g$ and $E_h \subseteq E_g$. H is the proposed secure netlist. H is the same as G except
that some edges are removed to be hidden in the BEOL layers. Both G and H are
directed acyclic graphs.

A vertex (gate) $u \in V_g$ is defined to be k-*secure* if it is indistinguishable from
at least $k - 1$ other vertices in V_g. Formally, u is k-secure if there exist k distinct
vertices $v_i \in V_g, i \in [1, k]$ and mappings $\phi_i : V_g \rightarrow V_h, i \in [1, k]$ such that
every ϕ_i is a subgraph isomorphism from G to H and for all $i \in [1, k], \phi_i(u) =
v_i$. See the definition of subgraph isomorphism in Sect. 4.2 where a satisfiability
based algorithm was presented to determine a mapping if one exists. Note that u
is indistinguishable from each v_i and u itself may be one of the v_i's. Each gate is
1-secure by definition. The maximum value of k is $|V_g|$.

H is a k-*secure* version of G provided every gate in V_g is at least k-secure. The
maximum k-security of H is the maximum k value for which every gate is k-secure.
Imeson et al. proved that determining whether a graph H is a k-secure version of G
is NP-complete [1].

Let $\sigma(G, H)$ denote the maximum k-security of $H \subseteq G$. Let $E' = E_g - E_h$
denote the set of edges lifted from G to obtain H. The cost of lifting, $c(G, E')$, is
proportional to $|E'|$ in general since the BEOL fab cost is assumed to be proportional
to the number of BEOL wires. Hence, the goal of lifting is to lift as few edges
as possible to obtain a specified security level k. That is, minimize $|E'|$ such that
$\sigma(G, H) \geq k$.

Imeson et al. proposed a greedy wire lifting heuristic to lift enough wires to
obtain a desired level of k-security. Although Imeson et al. used the context 3D ICs
to introduce this method, it is applicable to 2D and 2.5D designs as well.

5.1.2 Algorithm

A greedy heuristic can be used to determine the smallest set of edges to lift from
a netlist given a security level k. This is shown as the *lift_wires* procedure (line
7) in Algorithm 28. The procedure initially assumes that all edges in the graph
are marked for lifting to attain the best possible security (line 8). Then the edges

---

Algorithm 28: *Lift Wires* for k-Security [1]

Input: Netlist Graph $G = (V_g, E_g)$, Desired Security Level k
Output: Edges to Lifted E_R

1 $\{G_1, G_2, \ldots, G_P\} = \text{partition}(G)$;
2 $E_R = E - \cup_{i \in [1,P]} E_i$;
3 **for** $i \in [1, P]$ **do**
4 $\quad\mid\quad E_R = E_R \cup \text{lift\_wires}(G_i, k)$;
5 **end**
6 **return** E_R;

---

7 **lift_wires** *(G,k)*

Input: Netlist Graph $G = (V_g, E_g)$, Desired Security Level k
Output: Edges to be Lifted E_R

8   $E' = E_g$;
9   **while** $|E'| > 0$ **do**
10 $\quad\mid\quad s = 0$;
11 $\quad\mid\quad$ **foreach** $e \in E'$ **do**
12 $\quad\mid\quad\quad\mid\quad E' = E' - \{e\}$;
13 $\quad\mid\quad\quad\mid\quad$ **if** $\sigma(G, E') > s$ **then**
14 $\quad\mid\quad\quad\mid\quad\quad\mid\quad s = \sigma(G, E')$;
15 $\quad\mid\quad\quad\mid\quad\quad\mid\quad e_b = e$;
16 $\quad\mid\quad\quad\mid\quad$ **end**
17 $\quad\mid\quad\quad\mid\quad E' = E' \cup \{e\}$;
18 $\quad\mid\quad$ **end**
19 $\quad\mid\quad$ **if** $s < k$ **then**
20 $\quad\mid\quad\quad\mid\quad$ **return** E';
21 $\quad\mid\quad$ **end**
22 $\quad\mid\quad E' = E' - \{e_b\}$;
23 **end**
24 **return** E';

---

are progressively restored (line 22). In each iteration (of the while loop on line 9), the algorithm greedily identifies the best edge e_b to restore, that is, the edge that when restored causes the least reduction in security (lines 10–18) while yet ensuring that the security does not fall below the required k. If there is no such edge, then the current solution is returned (lines 19–21). To compute the maximum security σ of a solution, the satisfiability formulation discussed in Sect. 4.2 is used to repeatedly invoke a SAT solver while adding the previous solutions as constraints. Alternatively, additional constraints can be added to explicitly check whether a node $u \in V_g$ can be mapped to another node $v \in V_g$ and this can be repeated for all pairs of nodes. This process determines all possible mappings, that is, the maximum value of k is determined.

This greedy heuristic is suboptimal in general since the size of E' depends on the order of identifying and deleting the edges. Although efficient SAT solvers exist, satisfiability checking is an NP-complete problem. To scale the method for larger problems Imeson et al. suggested a simple technique [1] that is shown as the main procedure in Algorithm 28. In this technique, the netlist graph is split into P partitions such that each partition graph generates a tractable SAT problem, say, corresponding to 1000 gates. All inter-partition wires E_R, i.e. the cutset, are assumed to be lifted (line 2). In addition, the *lift_wires* procedure is used to identify intra-partition wires to be lifted from each partition (lines 3–5) while ensuring k-security.

5.1.3   Discussion

Imeson et al. [1] analyzed the efficacy and performance of the approach using the c432 circuit from the ISCAS-85 benchmark suite and a DES encryption circuit that has about 35,000 gates. Experiments with c432, which has about 200 gates, show that the greedy wire lifting heuristic provides a greater level of security compared with random wire lifting. For example, with 80 of the 303 wires left in FEOL, the random lifting results in 1-security (i.e. completely insecure circuit), while the greedy choice of wires results in 23-security. In this circuit, at least 47% of the wires need to be lifted using the *greedy heuristic* to obtain any meaningful security. This implies significant BEOL cost. However, as more wires are lifted beyond this, security increases rapidly, reaching 48-security when all wires are lifted.

Layout anonymization is achieved by first removing the wires marked for lifting and then placing and routing the remaining FEOL circuit. Lifted BEOL wires are separately routed. This results in a FEOL layout in which the proximity of gates reveals no connectivity information and the BEOL wires are evenly distributed across the die. When compared with the original c432 circuit, the 8-secure c432 3D circuit (about 66% of the lifted wires) has 1.6x power, 1.8x delay, and 3x area. The 48-secure circuit (all wires lifted) has 1.92x power, 2.1x delay, and about 5x area. Diverse libraries containing a more variety of gates allow a circuit to be optimized better but compromise security. For this circuit, when libraries with 3, 4, 5, 6, and 7 different gate types were used, the maximum attainable k-security levels were 48, 24, 13, 7, and 4, respectively.

For the DES circuit, using recursive partitioning and lifting about 70% of the wires, a 64-secure design could be obtained. This occupied 2.38x area is compared to the original 1-secure design. Detailed analysis of a potential attack scenario against the DES circuit shows that the attack cost in terms of both area and time would increase significantly.

k-Security is an important metric to characterize the level of security attained by a split design in the context of layout recognition attacks. k-Security is defined for acyclic graphs and is, therefore, readily applicable to combinational logic circuits. The greedy wire lifting method is also readily usable for combinational circuits. The metric and the wire lifting method may be adapted to sequential circuits as follows: First, lift all the wires connected to the flip-flops. This leaves only combinational logic without cycles. The k-security metric and the wire lifting procedure can then be applied to this circuit.

As noted, the greedy wire lifting method is *suboptimal*. A challenge for defense methods that aim to maximize security or ensure a desired level of security is to do so while minimizing the number of wires lifted (i.e. reduce the BEOL cost) and containing the PPA overhead. We will discuss a series of methods that aim to accomplish this goal.

5.2 Simultaneous Wire Lifting and Cell Insertion

Imeson et al.'s defense method, discussed in the previous section, considered lifting wires to increase the level of k-security. It is also possible to increase the k value by adding additional *dummy* cells to the design. Li et al. [2, 3] proposed a method based on simultaneous wire lifting and cell insertion to satisfy the k-security criterion. Recall that a circuit is k-secure if each cell is indistinguishable from at least $k - 1$ other cells in the netlist from an attacker's perspective. That is, the probability of the cell being correctly identified in the layout is $\leq 1/k$.

Imeson's greedy approach to wire lifting has several drawbacks: (1) Unique cells can always be identified no matter how many wires are lifted. (2) The method often leads to excessive wire lifting and increased wire lengths leading to increased BEOL fabrication cost and, possibly, excessive PPA overheads. (3) The method uses a satisfiability checker that is impractical to handle large circuits. Li's method overcomes all of these drawbacks by considering the addition of *dummy cells* and *dummy wires* to the original netlist.

5.2.1 Objective

Let $G = (V_g, E_g)$ be a directed graph representing the original netlist. Let $t_g :$ $V_g \rightarrow T$ denote the type of each cell where T is the set of all cell types allowed in the netlist. Let $w_g : V_g \rightarrow \{0, 1\}$ be a binary valued function denoting the choice of cells for protection. To protect a cell c, the user would set $w(c)$ to 1.

Let $H = (V_h, E_h)$ denote a directed graph representing the FEOL circuit generated from G. t_h and w_h are type and protection choice functions for H defined similar to t_g and w_g, respectively. H is initially generated as follows:

1. For each $v \in V_g$, add v' to V_h and set $t_h(v') = t_g(v)$ and $w_h(v') = w_g(v)$. We use a mapping function $b : V_g \rightarrow V_h$ to denote the nodes in V_h that correspond to nodes in V_g. We set $b(v) = v'$. We require that $t_g(v) = t_h(b(v))$.
2. For each edge $(u, v) \in E_g$, we add $(b(u), b(v))$ to E_h.

Consider the netlist graph shown in Fig. 5.1a. Each color indicates a type. Suppose all the nodes are marked for protection at least at level-2 security. Figure 5.1b shows an FEOL implementation in which of all the nets are lifted into BEOL to achieve 1-security for the overall graph and 2-security for 4 nodes, namely, 1',

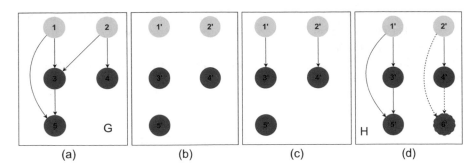

Fig. 5.1 Wire lifting and cell insertion (based on [2]). (**a**) A netlist graph, (**b**) 1-secure FEOL graph with all edges lifted, (**c**) 1-secure FEOL graph with reduced BEOL cost, (**d**) 2-secure FEOL graph with dummy nodes and edges added

2', 3', and 4'. Node 5 can be uniquely located by an attacker. Figure 5.1c shows the result of slightly better lifting to attain the same level of security with the BEOL cost reduced to three nets. Figure 5.1d shows a 2-secure FEOL implementation in which a dummy node 6' and two dummy edges to it are inserted with the BEOL cost reduced to one net. The increased FEOL gates and wires are expected to be absorbed in the available area or can be considered as the cost of increased security. Node 5 is no longer vulnerable.

To generate a protected FEOL circuit, the following three operations are postulated on H:

1. *Wire Lifting:* If edge $e \in E_h$ is lifted to BEOL then $E_h = E_h - \{e\}$ and V_h remains unchanged.
2. *Dummy Cell Insertion:* If a new cell represented by a new node v' and with cell type $t \in T$ is inserted, then $V_h = V_h \cup \{v'\}$, $t_h(v') = t$, $w_h(v') = 0$, and E_h remains unchanged.
3. *Dummy Wire Insertion:* If a new edge $e = (u', v')$, where v' is a dummy cell, is inserted, then $E_h = E_h \cup \{e\}$ with V_h unchanged. Only edges pointing to the dummy cells are allowed for insertion to ensure well-formed circuits.

The objective is to determine the best set of operations to perform, that is, to determine which wires to lift and which dummy cells/wires to insert to obtain a desired level of security.

5.2.2 Secure Implementations

To insert a trojan at node $v \in V_g$, the attacker needs to locate the corresponding node $b(v) \in V_h$. Imeson et al. [1] used graph isomorphism to establish this correspondence as discussed in Sects. 4.2 and 5.1.

G and H are isomorphic if $b : V_g \rightarrow V_h$ is a bijective mapping such that $(u, v) \in E_g$ if and only if $(b(u), b(v)) \in V_h$ and $t_g(u) = t_h(b(u))$ and $t_g(v) = t_h(b(v))$. If only wire lifting is considered to protect the circuit, then $V_h = V_g$ and $E_h \subseteq E_g$. There must be a subgraph of G that is isomorphic to H such that for each node in V_g, one or more nodes in V_h can be identified as its potential implementations. This was the definition used in Sects. 4.2 and 5.1.

Insertion of dummy nodes and edges necessitates a new approach to the definition of a suitable isomorphism relation. This is based on the following observations:

1. All nodes in G are retained in H. That is, $\forall v \in V_g, \exists v' \in V_h$ such that $b(v) = v'$.
2. Since dummy edges can only point to dummy cells, all edges incident on non-dummy cells in H should also exist in G. That is, if $v \in V_g$, $v' \in V_h$ such that $b(v) = v'$, then $\forall (u', v') \in E_h$ there must exist $u \in V_g$ such that $u' = b(u)$ and $(u, v) \in E_g$.

The concept of subgraph isomorphism is extended to accommodate these observations as follows.

A subgraph $G_s = (V_s, E_s)$ of $G = (V_g, E_g)$ is referred to as a *spanning subgraph* if $V_s = V_g$. G_s is said to be an *induced subgraph* if $\forall (u, v) \in E_g, (u, v) \in E_s$ if and only if $u, v \in V_s$. Given two graphs G and H, G is *spanning subgraph isomorphic* to H if there exists a spanning subgraph of G that is isomorphic to an induced subgraph of H. Spanning subgraph isomorphism captures the condition that needs to be satisfied for an attacker to find an implementation $b(v)$ in the layout of a target node v in the netlist. Note that, dummy edges can only be inserted pointing to dummy cells. Valid isomorphic relations should meet this restriction.

For example, G_s in Fig. 5.2.a shows a spanning subgraph of the netlist graph G in Fig. 5.1a. H_s in Fig. 5.2b is an induced subgraph of H shown in Fig. 5.1d. G_s is isomorphic to H_s under the mapping $b(1) = 2', b(2) = 1', b(3) = 4', b(4) = 3'$, and $b(5) = 6'$. Hence, G is spanning subgraph isomorphic to H.

Multiple spanning tree isomorphisms may exist between G and H. For a node in $v \in V_g$, let $C(v)$ be the set of candidate nodes in V_h. For each candidate node, the number of isomorphisms that map it to the original node may be different.

Fig. 5.2 Spanning and induced subgraphs (based on [2]). (**a**) Spanning subgraph of G, (**b**) Induced subgraph of H

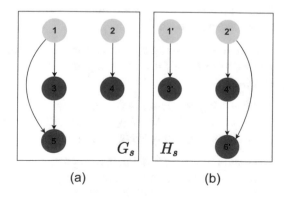

(a)   (b)

Candidates with a large number of possible ways for the isomorphic mapping are likely to be selected by the attacker as a potential site for trojan implantation for v. For candidate node $v' \in C(v)$, let $\mathcal{S}_v(v')$ denote the set of valid spanning subgraph isomorphisms that map v' to v. Then, the *probability of candidacy* of v' is defined as

$$\mathcal{P}_v(v') = \frac{|\mathcal{S}_v(v')|}{\sum_{u' \in C(v)} |\mathcal{S}_v(u')|} \tag{5.1}$$

Given the netlist graph G and layout graph H, a cell $v \in V_g$ is said to be a *k-secure cell* in H provided,

$$\sum_{u' \in C(v)} \mathcal{P}_v(u') \cdot w_h(u') \leq \frac{1}{k} \tag{5.2}$$

The layout graph H is a *k-secure implementation* of G provided $\forall v \in V_g$ with $w_g(v) = 1$, v is k-secure in H.

Checking for graph isomorphism is computationally intensive. Li et al. [2] proposed a heuristic solution based on the privacy preserving network publication [4]. The heuristic is based on the following definition: A graph is *k-isomorphic* if it consists of k disjoint *isomorphic subgraphs*.

Consider the graphs G and H in Fig. 5.1. H is 2-isomorphic with disjoint subgraphs H_1 and H_2 with node sets $V_{h,1} = \{1', 3', 5'\}$ and $V_{h,2} = \{2', 4', 6'\}$, respectively. 1' and 2' are said to be in the *same position* in H_1 and H_2, respectively. If $1' \in C(1)$, then $2' \in C(1)$. Further, $\mathcal{P}_1(1') = \mathcal{P}_1(2')$. If $b(1) = 1'$ and $w_h(2') = 0$, then node 1 is 2-secure. Based on this observation, the following lemma is proved in [2].

Given G and H, let H be k-isomorphic and let $\{H_1, H_1, \ldots, H_k\}$ be the k disjoint isomorphic subgraphs of H.

Lemma *A node $v \in V_g$ with $w_g(v) = 1$ and $b(v) \in V_{H_i}$, for some $i \in [1..k]$, is k-secure if all $u' \in V_{H_j}$, $j \neq i$, where u' and $b(v)$ are in the same position in H_j and H_i, respectively, satisfy $w_h(u') = 0$.*

Since only the nodes with a weight of 1 are required to be k-secure, the following theorem can be stated:

Theorem *G is k-secure with respect to H (equivalently, H is a k-secure implementation of G) if for all $v \in V_g$ with $w_g(v) = 1$, the following conditions are satisfied:*

1. *$b(v) \in V_{H_i}$ for some $i \in [1..k]$.*
2. *For all $u' \in V_{H_j}$, $1 \leq j \leq k$, $j \neq i$, where u' and $b(v)$ are in the same position in H_j and H_i respectively, $w_h(u') = 0$.*

Graph G in Fig. 5.3 is a netlist graph and H is an FEOL graph. H is divided into three subgraphs. H_1 and H_2 are isomorphic to each other and H_0 is the remaining portion of H. Nodes with a weight of 1 are circled in dark blue. These protected

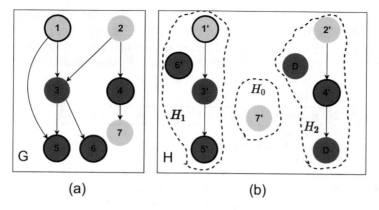

Fig. 5.3 2-security (based on [2]) (**a**) Netlist graph, (**b**) FEOL graph

nodes 1',4',5,6' are in either H_1 or H_2. Their "same place" counterparts, that is, 2',3',D,D, have zero weights. Hence, according to the previous lemma, they are 2-secure. Node 7 is unprotected and happened to be included in a separate subgraph named H_0. G is 2-secure with respect to H.

5.2.3   Netlist Manipulation Algorithm

The above theorem provides a heuristic for achieving k-security while allowing the insertion of dummy nodes and edges. A trivial method to generate a k-isomorphic implementation of G is to simply make k copies of G. However, this is wasteful of area, power, and possibly performance.

A constructive algorithm to generate the FEOL and BEOL layers is shown in Algorithm 29. Given the netlist G and the desired security level k, the algorithm generates from G a k-isomorphic graph H with isomorphic subgraphs $H_1, \ldots H_k$ such that all nodes with non-zero weights (i.e. protected nodes) are added to these subgraphs. V_{crit} is the set of critical nodes to be protected and V_r is the set of remaining nodes (lines 1–2). In each iteration (lines 3–17), k nodes of the same type are selected such that exactly one of those nodes is critical, and are inserted into the isomorphic (sub)graphs $H_1, \ldots H_k$ (line 14). To select the nodes, first, all the remaining nodes are clustered by their type (line 7) and k nodes with minimum cost are selected from each cluster using a MILP (Mixed Integer Linear Programming) formulation (line 8). The k nodes with the least cost among all the clusters are selected (lines 4–13) and inserted into the isomorphic subgraphs. At the end, H consists of the graph constructed out of the isomorphic subgraphs H_i, $1 \le i \le k$ plus an additional subgraph consisting of the remaining nodes $H_0 = V_r$ that were not included in any H_i.

For each cluster, nodes for the k subgraphs are selected using the procedure SelectNodes (line 8) that solves a MILP formulation of the selection problem subject to certain correctness constraints and a cost minimization goal. Using the variables in Table 5.1, the constraints and the optimization goal are explained below:

Algorithm 29: Iterative FEOL generation using MILP [3]

Input: Netlist G, Security Level k
Output: k-Isomorphic Layout Graph H
1 $V_{crit} = \{v \in V_g | w_g(v) = 1\}$;
2 $V_r = V_g$;
3 **while** $V_{crit} \neq \emptyset$ **do**
4 $\quad$ $V_{min} = \emptyset$;
5 $\quad$ $c_{min} = \infty$;
6 $\quad$ **for** *each cell type* $t \in T$ **do**
7 $\quad\quad$ $V_t = \{v \in V_r | t_g(v) = t\}$;
8 $\quad\quad$ $V_{sel}, c_{sel} = \text{SelectNodes}(k, V_t)$;
9 $\quad\quad$ **if** $c_{min} > c_{sel}$ **then**
10 $\quad\quad\quad$ $V_{min} = V_{sel}$;
11 $\quad\quad\quad$ $c_{min} = c_{sel}$;
12 $\quad\quad$ **end**
13 $\quad$ **end**
14 $\quad$ InsertToFEOL($V_{min}, H_1, H_2, \ldots, H_k$);
15 $\quad$ $V_{crit} = V_{crit} - V_{min}$;
16 $\quad$ $V_r = V_r - V_{min}$;
17 **end**
18 $H_0 = V_r$;
19 $H = \{H_0, H_1, H_2, \ldots, H_k\}$;
20 **return** H;

Table 5.1 Notation used in MILP formulation [3]

Variables	Definitions
x_i	= 1 if i-th node is selected; = 0 otherwise
x_{ij}	= 1 if i-th node is inserted in H_j
w_i	weight $w_h(i)$ of the i-th node
d_j	= 1 if a dummy node is inserted in H_j
y_l	= 1 if an edge can be added from the l-th location to the current location in $H_1, H_2, \ldots, H_k$
y_{lj}	= 1 if an edge can be added from the l-th location to the current location in H_j
z_l	= 1 if an edge can be added from the current location to the l-th location in $H_1, H_2, \ldots, H_k$
z_{lj}	= 1 if an edge can be added from the current location to the l-th location in H_j
IN_{ij}	set of starting locations of edges pointing to the current location that can be added if the i-th node is added to H_j
OUT_{ij}	set of ending locations of edges pointing from the current location that can be added if the i-th node is added to H_j
RES_i	set of edges connected to the i-th node from the unadded node

1. Each node i can be inserted into at most one subgraph H_j:

$$\sum_{j=1}^{k} x_{ij} = x_i, \quad \forall i \tag{5.3}$$

2. For each subgraph H_j, exactly one node (either real or dummy) should be inserted in H_j:

$$\sum_{i} x_{ij} + d_j = 1, \quad \forall j \in \{1, 2 \ldots, k\} \tag{5.4}$$

3. To satisfy the theorem, exactly one of the nodes being inserted should be critical:

$$\sum_{i} x_i w_i = 1 \tag{5.5}$$

4. In any H_j, an edge pointing from the l-th position to the current position can be added in the FEOL layer provided one of these two conditions is met: (1) a dummy cell is inserted in the current position or (2) node i with $l \in \text{IN}_{ij}$ is inserted. In addition, to satisfy subgraph isomorphism, an edge from the l-th position should also be added in all subgraphs.[1]

$$y_{lj} \le \sum_{i} x_{ij} \cdot 1_{l \in \text{IN}_{ij}} + d_j, \quad \forall j, l$$

$$y_l \le y_{lj} \qquad\qquad\qquad \forall j, l$$

These two inequalities can be simplified by substitution as

$$y_l \le \sum_{i} x_{ij} \cdot 1_{l \in \text{IN}_{ij}} + d_j, \quad \forall j, l \tag{5.6}$$

Similarly, in any H_j, an edge from the current position pointing to the l-th position can be added in the FEOL layer provided node i with $l \in \text{OUT}_{ij}$ is inserted. In addition, to satisfy subgraph isomorphism, an edge to the l-th position should also be added in all subgraphs.

$$z_{lj} \le \sum_{i} x_{ij} \cdot 1_{l \in \text{OUT}_{ij}}, \quad \forall j, l$$

$$z_l \le z_{lj}, \qquad\qquad\qquad \forall j, l$$

These two inequalities can be simplified by substitution as

---

[1] 1_B denotes the identity function that equals 1 when the Boolean valued expression B evaluates to true; 0 otherwise.

$$z_l \leq \sum_i x_{ij} \cdot 1_{l \in \text{OUT}_{ij}}, \qquad \forall j, l \qquad (5.7)$$

5. While satisfying these constraints, the objective is to select nodes that would minimize a weighted cost function consisting of three components: (1) the number of BEOL edges, (2) the number of edges that can be added back to the FEOL layers, and (3) the combined area of all the dummy nodes inserted. This is expressed as

$$\underset{x,d}{MIN} \; \alpha \sum_i |\text{RES}_i| \cdot x_i - \beta k \sum_l (y_l + z_l) + \gamma A \sum_j d_j \qquad (5.8)$$

where A represents the area of each dummy node and α, β, and γ are the weighing factors to explore the tradeoff between dummy node insertions and wire lifting. Costs of wire lifting and dummy nodes are assumed to be independent of each other to allow the linear dependence in the optimization goal.

MILP solvers are inefficient for large problem sizes. To increase the computational efficiency, a Lagrangian relaxation (LR) method is suggested in [3] by relaxing y_l and z_l to be continuous variables without changing the optimal solution. By further transforming one of the resulting subproblems into a network flow problem for which efficient algorithms exist, Li et al. demonstrated significant computational speed-up.

5.2.4   Secure Layout

While Algorithm 29 generates a k-secure design that can defeat satisfiability based layout recognition attacks, the resulting layout may be vulnerable to proximity based attacks since standard placement tools tend to place connected cells close to each other. Additional care should be taken during layout generation to defeat proximity based recovery. Li et al. also proposed a layout generation method by introducing virtual nets that enforce additional placement constraints to bring close together cells that may or may not be connected in the original netlist.

Let $H_1, \ldots H_k$ denote the k-isomorphic subgraphs of H. Note that H may contain additional nodes that are not included in any of these subgraphs. Denote the subgraph consisting of these nodes by H_0 (line 18 of Algorithm 29). Consider edge $(u, v) \in E_g$ such that $u' = b(u) \in H_i$ and $v' = b(v) \in H_j$ where $i, j \in [0..k]$.

1. If $i = j = 0$, then (u', v') exists in the FEOL layers. Hence, no virtual net is added.
2. If $i = 0$ and $j \in [1..k]$, then (u', v') is lifted to the BEOL part. In this case, $\forall v'_n \in V_{H_n}$ with v'_n in the same position as v' and $1 \leq n \leq k$, a virtual net (u', v'_n) is inserted.

3. If $j = 0$ and $i \in [1..k]$, then (u', v') is lifted to the BEOL part. In this case, $\forall u'_n \in V_{H_n}$ with u'_n in the same position as u' and $1 \le n \le k$, a virtual net (u'_n, v') is inserted.
4. If $i \ne 0$, $j \ne 0$, and $i = j$, then, $\forall u'_n, v'_n \in V_{H_n}$ with u'_n and v'_n in the same positions as u' and v', respectively, and $1 \le n \le k$, a virtual net (u'_n, v'_n) is inserted.
5. If $i \ne 0$, $j \ne 0$, and $i \ne j$, then no virtual nets are inserted.

5.2.5 Discussion

Li et al. [3] evaluated the simultaneous wire lifting and cell insertion method using the selected ISCAS-85 benchmarks and three other functional circuits with the number of nodes from 214 to 5720. Switching probabilities of the signals are computed and rarely switching nodes are selected for protection. Layers M1-M3 are assumed to be in FEOL and M4 and above in BEOL.

When 5% of the nodes are protected with $k = 10$, $\alpha = 0.5$, $\beta = 2$, and $\gamma = 0.6$, the MILP based algorithm takes a few seconds for small benchmarks (under 500 nodes), while the greedy wire lifting method [1] discussed in Sect. 5.1 took several hours. On larger benchmarks, the greedy method timed out, while the MILP method took under an hour for the largest benchmark attempted. The LR method with 10 iterations achieves a further speed-up of 9.9x over the basic MILP method. In addition, as the number of protected nodes increased, the LR method is more scalable. With 18% of the nodes protected, the LR method completed in about 4 min, while the basic MILP method timed out. The MILP algorithm introduces, on average, 104% less wire length overhead and 3.97% less area overhead compared with the greedy method. Even better reductions were observed as k value was increased with the greedy wire lifting method unable to handle $k > 15$.

With the addition of virtual nets that lead to compact placement, Li's method achieved 97.5% wire length overhead reduction. Distance distributions showed that the distances between protected nodes and their candidate nodes are similar. This indicates that strict proximity based attacks are unlikely to be successful. If the closest candidate pin was selected as the matching pin, then essentially no matching pairs could be found.

For large k values, both area and wire length overheads increase significantly. Power overhead was not measured. Although Li et al.'s method was demonstrated on relatively small examples, the results show that this method is significantly better than the greedy wire lifting method. Similar to the greedy method, Li et al.'s method can be directly applied only to combinational logic circuits.

5.3   Combined Layout Camouflaging and SM

So far, we have discussed SM as the only method employed to improve security from some supply chain attacks. Other methods such as *layout camouflaging* and logic obfuscation were also proposed to defend against various attacks in the supply chain. In this and the next few sections, we will discuss defense methods employing SM along with other Design for Trust (DfT) methods to enhance the security of a design.

Layout camouflaging [5] refers to the addition of dummy contacts or vias to defend against reverse engineering attacks. However, previous camouflaging methods resulted in significant layout cost and hence were applied to selected gates across the layout in the FEOL layers. This results in limited security and vulnerability against other attacks. In particular, designs with inadequate camouflaging can be successfully reverse engineered using the SAT attack discussed in Sect. 4.1.3. Further, the camouflaged FEOL layouts can only be fabricated at a secure foundry. Patnaik et al. [6, 7] proposed integrating layout camouflaging methods in BEOL layers along with split fabrication. They have suggested a low-overhead method for full-chip camouflaging to obfuscate connections in the BEOL layers. This provides defense against the SAT attack. In addition, it allows split fabrication of the FEOL part at an untrusted foundry. Patnaik et al.'s method is described in this section.

5.3.1   Objective

Several researchers have proposed obfuscating the interconnects by using a mix of conducting vias and non-conducting vias using materials that could defeat delayering attempts. For example, Chen et al. [8] suggested the use of magnesium (Mg) for conducting vias and magnesium oxide (MgO) for *dummy vias*. When exposed during delayering Mg vias oxidize into MgO within minutes and become indistinguishable from the dummy vias. Other emerging interconnect technologies may make BEOL via obfuscation even more effective [9].

Patnaik et al. [6] proposed a general *obfuscating cell* illustrated in Fig. 5.4a. In this primitive, an input wire of a cell is camouflaged with $n - 1$ other unrelated wires that may be randomly selected. The $n - 1$ wires act as dummy substitutes for the real input. A specific, cost-effective rendition of the scheme is shown in Fig. 5.4b. For an arbitrary two-input gate, each input (shown in green) is camouflaged with an unrelated wire (shown in red) and constants 0 and 1. By selecting one of the possible signals on each input, it is possible to realize up to 16 functions one of which is the real one and the rest are dummies. Figure 5.4c shows the conceptual implementation in which M6 and above are in the BEOL. Dummy vias are marked by red crosses and the real via is marked by a green circle. Figure 5.4d shows a possible layout view of the camouflaging cell in which A is the real input pin, B, C, and D are the

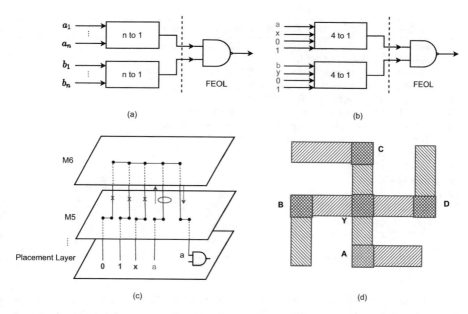

Fig. 5.4 Obfuscating cell (based on [6]) (**a**) Generic obfuscating cell, (**b**) 4 to 1 obfuscating cell, (**c**) Conceptual implementation, (**d**) Possible layout view

dummy input pins, and Y is the real output pin. Layer assignments are not shown, but variations (M5/M6 or M3/M4) can be implemented and used.

Constant values 0 and 1 are realized using special standard cells called TIE (tie high, tie low) cells. Preponderance of these cells in the layout may alert an attacker to simply disregard them. Further, the attacker may become aware of the implausible functionalities due to the addition of the fixed values and try to eliminate them. Patnaik et al. proposed to transform the netlist for a deeper integration of the fixed values as follows: (1) transform selected inverter and buffer cells into camouflaged cells of other types such as NAND, NOR, etc; (2) insert additional camouflaged gates of randomly selected types to fill randomly selected white space. These gates drive the dummy inputs of real cells in the vicinity so that they cannot be removed for lack of connectivity.

5.3.2 Tool Flow

The tool flow for the camouflaging methodology is shown in Algorithm 30. *Camouflaging scale* is defined as the ratio of the number of obfuscated nets to the total number of nets in the design. *Metal layer* ratio is the ratio of lower (say, M3 to M4) to higher layers (say, M5 to M6) where camouflaging should be incorporated. Following initial synthesis, place, and route, netlist transformation and *TIE cell*

Algorithm 30: Design flow for camouflaging BEOL interconnects [7]

    Input: Netlist N, Camouflaging Scale s, Metal Layer Ratio m
    Output: Camouflaged Layout L
1 Synthesize, place and route N to generate layout L;
2 Transform selected inverter and buffer cells in L;
3 Add regular and camouflaged TIE cells to L;
4 Embed obfuscating cells in L for selected gates according to s and m;
5 Select dummy nets for inputs of gates while avoiding combinational loops;
6 Perform trial routing of L and evaluate congestion;
7 **if** *congested* **then**
8 | Ask user to revise m;
9 | **go to** 4;
10 **end**
11 Optimize ECO, legalize and re-route L;
12 Perform design closure;
13 Remove obfuscating cells from L;
14 **return** L;

---

insertion are performed as discussed above (lines 1–3). Next, $s\%$ of nets are selected and $m\%$ are assigned to lower BEOL layers and the rest to upper BEOL layers for camouflaging. Both inputs of each selected gate are marked for obfuscation. Obfuscating cells are then inserted (line 4). These cells facilitate placement of dummy vias and routing of real and dummy wires; they have no baring on the FEOL layers. Dummy nets are selected from nearby nets while avoiding combinational loops (line 5). Following trial routing, if excessive congestion is noticed, then the user will be asked to revise m or enlarge the die area and the process is repeated (lines 6–10). Following completion of routing and design closure, the obfuscating cells are removed while leaving behind the real/dummy vias and wires (lines 11–13). Layouts of obfuscating cells and physical design methods are discussed in [7].

5.3.3   Discussion

Patnaik et al. [6] have reported extensive experimental results using the ISCAS-85, ITC-99, and EPFL [10] suites of benchmarks using up to 10 layers. Obfuscation cells were implemented in two versions M3/M4 and M5/M6, the latter being the default. 50% of all the INV/BUF cells were transformed into camouflaged gates.

    Area overhead for the proposed method (measured in terms of the die outline necessary to achieve DRC-clean layouts) varied from 1.59% to 27.89% as s varied from 20% to 100%. Power overhead varied from 7.8% to 31.02% and delay overhead from 8.83% to 23.11%. Comparisons with a number of other layout camouflaging methods show that the proposed method is significantly more PPA-efficient.

Experiments with attacking camouflaged circuits using the SAT attack showed that the SAT attack becomes prohibitively expensive (timed out even after 7 days) once s crosses 40% for all benchmarks except one and for much lower s values for some benchmarks.

Network flow attack (Sect. 2.5) mounted on camouflaged circuits split after M3 showed a reduction in CCR by 7.48x when compared with the original circuits. Split after M4 resulted in a CCR reduction by 2.41x. The number of cut wires increased on average by about 6x and 19x when splitting after M3 and M4, respectively. For both splits, the HD values for the recovered netlists ranged over 40.71%–49.79% depending on s that ranged from 20% to 100%. More significantly, OER remained at 99.99% in all cases. Thus, the proposed method offers significant defense against both SAT and network flow attacks at a modest PPA cost.

When the crouting based proximity attack [11] (Sect. 2.4) was used against the 100% of camouflaged circuits, the number of vpins increased, on average, by 12.55x. E[LS] showed a similar increase. Figure of merit, FOM, increased by 9.30x and 24.15x for split layers M3 and M4, respectively. This shows that this method is resilient against crouting based proximity as well.

Patnaik et al.'s method is the first *full-chip camouflaging* method. This method defends against both SAT attacks and design constraint attacks. It is noteworthy that the power of the methods comes from combining the idea of camouflaging with the SM process. For the method to become practical, the technology assumed for obfuscating the vias should become accessible.

5.4   Combined Logic Encryption and SM

Recall that logic encryption, discussed in Sect. 4.1.2, is intended to prevent attacks by the end users. The correct key bits are stored in a secure memory. Sengupta et al. [12] proposed a method to use logic encryption to lock the FEOL circuit, hard-wire the correct key constants in the FEOL circuit but obfuscate the connectivity between the constant key values and the key gates in the BEOL layers. Hence, while the circuit functionality is locked in the FEOL part, it is unlocked in the BEOL part. This method protects the design from the usual FEOL foundry-based attacks against SM but does not necessarily protect against the end-user attacks since the key bits are not secured once the FEOL and BEOL parts are combined into the final IC at a secure foundry.

5.4.1   Methodology and Tool Flow

In Sengupta et al.'s method, the FEOL circuit is first locked using any logic encryption method by introducing key gates controlled by key bits. The constant key bit values are implemented by TIE (tie-to-1, tie-to-0) cells that are added to

---

Algorithm 31: Design flow for combined logic encryption and SM [12]

Input: Netlist N
Output: Split Layout L
`// Partitioning, synthesis, and logic locking`
1 Partition N hierarchically into a set of balanced partitions;
2 Synthesize, perform logic encryption for each partition, resynthesize, include TIE cells in place of key bits, add set_dont_touch constraints on TIE cells and set_dont_touch_network constraints on key nets, and merge the locked netlists;
3 Insert TIE cells in place of key bits;
   `// Layout synthesis`
4 Randomize the placement of TIE cells and fix their positions by setting set_dont_touch constraints;
5 Detach TIE cells from key gates;
6 Place the remaining cells in the design netlist;
7 Attach TIE cells back to key gates;
8 Performing routing;
9 Define stacked via constraints, lift key nets to designated BEOL layers and reroute in ECO mode;
10 Generate split layout L at the designated split layer;
11 **return** L;

---

the design. In the placement stage, these TIE cells are randomly placed and fixed in those random positions. The TIE cells are then detached from the key gates to avoid introducing any proximity-type hints and the remaining cells in the design are then placed. Then, the TIE cells are attached back to the key gates and routing is completed. ECO routing, after designating the key nets as new nets, is performed to lift key nets (nets connecting the TIE cells to key gates) into the designated BEOL layers and reroute the other nets as needed. Stacked vias are used at both ends of the key nets such that whole key nets are lifted at once and can be configured to different split layers. A proof of security guarantee against the usual FEOL foundry-based attacks against SM is presented in [12]. While the method is potentially vulnerable against oracle-guided attacks such as the SAT attack, it is assumed that an oracle is unavailable.

While any logic encryption method can be used, Sengupta et al. adapted an ATPG-based logic encryption method proposed in [13]. To reduce the computational cost, they proposed to partition the design using random but balanced hierarchical partitioning. Partitioning allows protection of all parts of the design and the ability to process the partitions in parallel. The overall tool flow is shown in Algorithm 31.

5.4.2 Discussion

Sengupta et al. [12] evaluated the proposed method using the ISCAS-85 and ITC-99 benchmarks using 128 key bits for logic encryption.

Their experimental studies show that when the ITC-99 circuits were secured using the proposed method, the network flow attack (Sect. 2.5, [14]) was able to recover only about 50% of the key nets logically correctly, i.e. connected to some TIE cell of the correct key bit value, regardless of the split level. This is no better than a random guess. In addition, physical CCR, which is the correct connectivity rate of the correct physical TIE cells, was in the range 0% to 2%, indicating the high level of physically secure key obfuscation. The average HD when the split is at M4 (M6) is 53% (25%). HD decreases when split level is higher since most of the design is available in FEOL. OER in all cases was 100% again indicating the high level of protection. Even when all the regular design nets are *assumed* to be correctly recovered, OER remains at 100% even after applying 10^6 random key-net guesses. Run-times for ITC-99 benchmarks varied from 5 to 18 h with most of the time consumed in the logic re-synthesis runs.

Additional experiments using the ISCAS-85 benchmarks to compare the proposed method with the routing perturbation (Sect. 3.8, [15]), concerted wire lifting (Sect. 3.9, [16]), and netlist randomization (Sect. 3.12, [17]) methods show that the proposed method is competitive or superior based on the PNR, CCR, HD, and OER metrics.

Note that this method is not designed as a defense against the SAT attack since the key values are not included in a secure memory. However, the idea of logic encryption is effectively used to keep the FEOL "locked" in the absence of any knowledge of the BEOL nets. If an attacker at the FEOL foundry has access to an oracle, then an oracle-guided SAT attack could be mounted using the methods discussed in Chap. 4.

5.5   Combined Layout Camouflaging and SM for 3D ICs

As discussed in Chap. 1, SM fits well with the 3D integration process and does so with *minimal cost overhead*. In addition, the redistribution layers in 3D ICs can also be effectively used for layout camouflaging. Patnaik et al. [18, 19] proposed a methodology and tool flow for incorporating layout camouflaging (LC) into split manufactured face-to-face (F2F) 3D ICs. While conceptually similar to the approach discussed for 2D ICs in Sect. 5.3, the threat model in this case consists of several potentially untrusted foundries. Different IP modules may be fabricated at different untrusted foundries. It is assumed that the redistribution layers (RDL) are manufactured at a trusted BEOL fabrication and stacking facility. As in the case of 2D camouflaging, Mg/MgO vias (or similar) are assumed for obfuscating the vertical 3D interconnects.

Algorithm 32: Tool flow for IP protection in 3D ICs [18, 19]

Input: Netlist N, 2D Layout L
Output: Camouflaged 3D Layout L_{3D}
  `// Design Partitioning`
1 Obtain timing data from 2D layout;
2 (N_t, N_b) = Partition N into a top part and a bottom part;
  `// F2F Via Planning`
3 Place N_b and determine F2F via locations;
4 Place F2F vias, insert switch boxes and perform on-track legalization;
5 Map via locations to top tier;
  `// Design Closure`
6 Place and route N_b and RDL;
7 Place and route N_t;
8 Encapsulate both partitions in wrapper;
9 Assemble and implement the design;
10 Annotate F2F via parasitics in the wrapper;
11 Perform PPA analysis and final checks;
12 **return** L_{3D};

5.5.1   Design Methodology for IP Protection

Tool flow for IP protection proposed by Patnaik et al. [18, 19] is shown in Algorithm 32. The flow begins with a baseline 2D layout and the corresponding netlist. The netlist is partitioned into two parts, *top* and *bottom*, and IO ports are created at the F2F interfaces between the two parts. Multiple partitioning methods are contemplated: (1) *Random partitioning.* (2) *Maximal cut-size partitioning.* In this case, in each timing path, gates are alternatively assigned to the top and bottom tiers. This is the worst case for PPA but maximizes security. The number of hidden BEOL nets is maximized. (3) *Timing-aware partitioning.* In this case timing slack for each gate is determined. Given a user-specified threshold, critical gates are assigned to the bottom tier and the other gates are assigned to the top tier. Timing slacks are recomputed and the procedure is repeated until even utilization is achieved for both layers. This is the recommended strategy for the best PPA vs. security tradeoff. (4) *Hierarchical partitioning.* If the design has hierarchies at the top level, then modules with a large degree of connectivity across the tiers are first separated and the other modules are distributed to balance the utilization of both tiers.

Aligning the ports at the bottom and top tiers can introduce a vulnerability that can be exploited if an attacker has access to both parts. To improve security, additional F2F ports are placed randomly in the top RDL and routed through the RDLs toward the original F2F ports at the bottom tier. Connectivity in the RDLs is further obfuscated using custom *switch boxes* that allow hidden mapping of four drivers to four sinks. Switch box uses Mg/MgO vias to camouflage the mapping. Pins of the switch box, representing F2F ports, are aligned with routing tracks. Figure 5.5a shows RDL randomization and the obfuscated switch boxes and

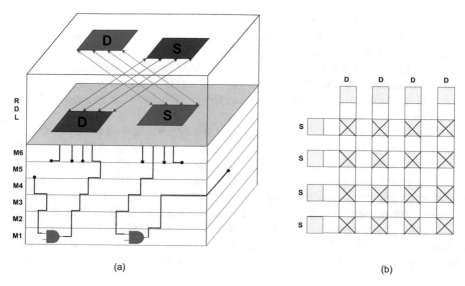

Fig. 5.5 Obfuscated switch box (based on [19]) (**a**) RDL randomization with switch boxes, (**b**) Conceptual layout view of a switch box

Fig. 5.5b shows a conceptual layout of the switch box in which each D (drive) pin can be connected to its corresponding real S (sink) pin through a real via. Three vias in each row are dummy vias. For legalization, each F2F port is moved into the core boundary. Actual placement is done on the closest, unoccupied on-track location. Following this, both tiers are placed and routed independently. No cross-optimization is performed between the tiers to allow anonymization. Bottom tier and the RDL with switch boxes are routed together. The partitions are encapsulated in a wrapper netlist and standard closure steps are performed before generating separate definition files for both the tiers and the RDL.

5.5.2   Discussion

Patnaik et al. [19] reported extensive experimental results. They have used 2D tools to model the 3D design flow using 6 metal layers for baseline 2D designs, 6 layers each for the top and bottom F2F chips, and 4 duplicated M8 layers for the RDL. Two RDL layers were used for the switch boxes and two for randomizing the routing. They have used benchmarks from the ISCAS-85 and ITC-99 sets.

With the random partitioning approach (without switch boxes and randomization), moving 50% of the gates, they have reported performance overheads in the range −3% to 18% and power overheads in the range 0% to 18%. Performance improvement in some cases was due to the reduction in wire lengths due to the vertical interconnect. With the maximal cut-size approach including switch boxes and randomizing the F2F ports, they have reported up to 60% overheads for the larger benchmarks. With timing-aware partitioning including switch boxes and port randomization, the performance overhead was in the range -8% to 5% and power overhead was in the range -3% to 15%. They have reported additional results for two SoC benchmarks, DARPA CEP [20] and a JPEG from opencores.org [21]. They have also reported better results compared to several previously proposed LC and SM schemes.

Patnaik et al. evaluated the effectiveness of the tool flow against the SAT attack as well as a proximity attack specifically developed for split fabricated 3D ICs [19, 22]. This attack attempts to recover the connectivities among the F2F ports between the tiers. It exploits unique mappings, if any, between the F2F sources and sinks, and proximity based layout hints while avoiding combinational cycles. The attack is much more efficient than the SAT attack discussed in Sect. 4.1.3. However, the 3D proximity attack results in PNR 20%–55%, HD 35%–50%, and CCR 20%–36%. These results indicate that while the attack can correctly recover some connections, the design functionality is still substantially obfuscated.

5.5.3   Design Methodology for Trojan Prevention

Patnaik et al. [19] also proposed a methodology and tool flow for preventing trojan insertion based on layout recognition attacks introduced in Sect. 1.12). In contrast to [1, 3], Patnaik et al. propose security-driven synthesis to ensure that k-security is incorporated during synthesis. To achieve k-security, isomorphic instances of vulnerable structures are introduced. The overall synthesis flow proposed is shown in Algorithm 33.

Given an RTL model, initial synthesis is performed to produce a baseline netlist. A designer then identifies, using any vulnerability analysis method [1, 16], gates and small subcircuits that are vulnerable as trojan implantation sites. These structures are

---

Algorithm 33: Synthesis flow for Trojan prevention in 3D ICs [19]

Input: RTL Design R, Security Level k
Output: Secure Gate Level Netlist N
1 Perform regular synthesis to obtain a baseline netlist N;
2 Perform vulnerability analysis to identify vulnerable gates and vulnerable structures;
3 Generate custom cells from the vulnerable structures;
4 **while** *security level is not attained and layout cost is acceptable* **do**
5 | Perform re-synthesis to obtain a new netlist N;
6 | Set "don't touch" constraints on the structures in the netlist;
7 | Decompose structures into constituent gates;
8 **end**
9 **return** N;

---

then defined as *custom cells* for synthesis. However, library characterization is not performed since they will be transformed back into their original form. Instead, data from the existing "similar" cells from the library (e.g. simple cells with a matching number of inputs) is utilized. Synthesis is repeated with these custom cells added to the library so that the synthesis tool uses these cells throughout the design. Once these cells are introduced, their input/output wires are demarcated for lifting into the BEOL RDL to ensure k-security. "Don't touch" directives are set on the synthesis-generated structure instance to avoid those gates being optimized. Based on the PPA cost and desired k-security level, the designer performs several iterations of synthesis.

Once the synthesis iterations are completed, the resulting 2D netlist is further processed to obtain a 3D layout using the tool flow discussed in Algorithm 32 with a few changes. Camouflaging of the RDL is not used. Randomization of F2F ports is performed but the use of obfuscated switch boxes is omitted. Input/output wires are lifted to RDL using a customized lifting process adapted from [16] to ensure k-security. Partitioning into top and bottom tiers is constrained to ensure that all the gates in decomposed instances of the isomorphic structures stay together. During timing-driven partitioning, each critical path without any isomorphic structures is allocated to one tier. Non-critical paths or paths with some isomorphic instances are randomly partitioned. Thus, the attacker cannot track the netlist structures to the isomorphic instances in either of the tiers.

5.5.4 Discussion

Patnaik et al. [19] have used the same setup discussed in the previous section except that two M8 layers were used for RDL. Synthesis run-times for the large IBM superblue benchmarks took about 4 h per iteration.

When 10% of the gates were identified to be vulnerable, the proposed method, after 5 iterations, resulted in, on average, 36.5% of the netlists covered by isomorphic cell instances. While this is stronger protection than the 10% sought, some of the vulnerable gates may not be protected in the final design due to re-synthesis. k-security levels achieved ranged from 30 to 1221. The area, power, and delay overheads for the final 2D designs compared to 2D baselines ranged from 16% to 154%, 41% to 64%, and 55% to 88% respectively.

When converted into 3D, using the modified 3D flow, the area (overall outline), power, and delay overheads compared to 2D baselines ranged from -40% to -25%, 8% to 46%, and 37% to 62%, respectively. k-security (measured as the sum of the least occurring structures in the two tiers) ranged from 30 to 400. As expected, the SAT attack (Sect. 4.1.3) to recover the missing RDL connections timed out after 72 h on all benchmarks. Even if the SAT attack succeeds to return a functionally correct circuit, it cannot return the structure that is necessary for trojan implantation.

5.6   Obfuscated Built-In Self-Authentication

If there is vacant space available in the layout, attackers can use that space to insert trojans. Filler cells can be used to fill all the vacant space to dissuade trojan insertions. However, it is possible for the attackers to recognize and remove the filler cells. Hence, the defense method should detect if any filler cell is removed. Xiao et al. [23, 24] proposed a systematic method to insert filler cells such that the removal of any filler cell can be detected. The method, named *Built-In Self-Authentication* (BISA), constructs fully testable modules using the filler cells and uses built-in self-test (BIST) methods to confirm that they are not deleted.

    Figure 5.6 shows the BISA circuit in which the filler cells are connected into multiple blocks where each block is structured as a tree ensuring complete testability. The blocks together form a circuit under test (CUT) that is tested by vectors produced using a test pattern generator (TPG) and evaluated using an output response analyzer (ORA). A linear feedback shift register (LFSR) is used as a shared TPG and a multiple input signature register is used as the ORA. BISA circuitry can be in the idle mode or test mode. In the idle mode the LFSR contains an idle pattern that is nominally set to alternating 0's and 1's. In the test mode, the LFSR is operated to produce test vectors to test the BISA blocks.

    ICs incorporating the BISA method could still be vulnerable to "redesign" attacks in which an attacker who recognizes the BISA cells can redesign portions of the circuit to make room for trojan cells. BISA method cannot preclude reverse engineering and cloning. However, SM can defend against redesign attacks as well as IP piracy and IC cloning. Xiao et al. [25] and Shi et al. [26–28] proposed incorporating BISA into split fabricated ICs in order to benefit from both techniques. While SM protects against targeted trojan insertions that require the attacker to locate specific logic structures in the layout, it still leaves the layout vulnerable to *untargeted trojan insertion* that only requires locating rare triggering nodes that are not detected by normal test procedures. BISA protects against such untargeted trojan insertions. The combined method is named *Ofuscated BISA* (OBISA). Two OBISA methodologies are proposed based on security vs. computational cost tradeoff during the BEOL net lifting step.

Fig. 5.6 Built-In self-authentication (BISA) architecture (based on [23])

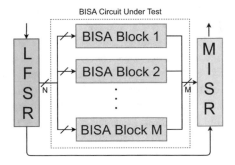

5.6.1   OBISA Obfuscated Connection Approach

OBISA obfuscated connection methodology assumes a simple SM approach: all nets from a certain metal layer, say M4, and above are delegated to BEOL. No other selection criterion is used. The overall methodology is shown in Algorithm 34. After initial layout design, OBISA cells are inserted in the unused space and are connected to form OBISA blocks that are tree-like logic structures. OBISA cells are standard cells that are used to fill the available placement slots while maximizing the variety of cells inserted so as to thwart cell composition analysis (Sect. 2.1). Following this, two types of nets are introduced:

- *Fanout nets connecting OBISA blocks:* A fanout net is a net inserted to drive an input of an OBISA block from a net in another OBISA block. These nets make it more difficult for the attacker to identify the OBISA blocks and are created based on the following rule: A net n in block i can drive an input of block j, $i \neq j$, provided the output of block j and n have no common related inputs from the LFSR. Related inputs are the bits in the LFSR that are in the input cones of these signals. This rule ensures that the fanout nets do not create redundancies that could compromise testability of the OBISA blocks.
- *Obfuscation connections:* An obfuscation net is a connection inserted to drive an input of an OBISA block from a net in the original circuit. This allows obfuscation of the logic gates in the original circuit with the OBISA cells, which makes it harder for the attacker to identify specific logic structures. Obfuscation nets are created using the following rule: A net n in the original circuit can drive the input i of an OBISA leaf cell g provided (1) the control value[2] of g is the same as the idle state value in the LFSR register bit driving g so that the activity in the original circuit does not propagate into the OBISA block when it is in the idle mode causing unnecessary power draw and (2) n has sufficient timing slack such that the added capacitance of i does not cause timing failures in the original circuit.

The net lifting step that identifies the BEOL nets is kept simple. In addition to all the nets above a selected metal level, say, M4, in the original circuit, all the feedback nets in the LFSR/MISR, nets connecting the flip-flops in the LFSR/MISR, and the OBISA mode select net are lifted to BEOL.

5.6.2   OBISA Wire Lifting Approach

Shi et al. [26, 28] proposed an alternative approach to OBISA incorporation named the *OBISA wire lifting* approach. In this approach, optimal selection of wire lifting

---

[2]Control value of AND/NAND gates is 0 and that of OR/NOR gates is 1 [29].

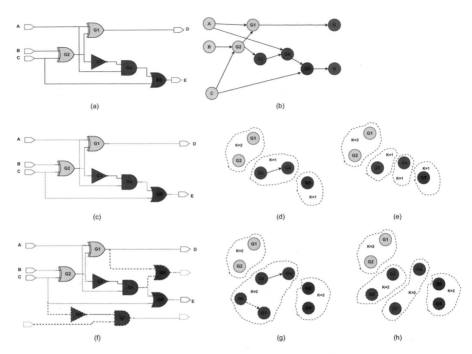

Fig. 5.7 OBISA for k-security (based on [28]). (**a**) Gate level design, (**b**) Graph representation, (**c**) Lifted nets shown in dotted lines, (**d**) FEOL graph, (**e**) FEOL graph when all nets are lifted, (**f**) Gate level design with added OBISA cells, (**g**) FEOL graph with selected nets lifted, (**h**) FEOL graph with all the nets lifted

is considered to obtain better security at the expense of computation time. The k-security metric discussed in Sect. 1.12 is used to quantify the level of security.

Consider the netlist in Fig. 5.7a. It is represented by a graph in Fig. 5.7b. Figure 5.7c shows the netlist with the dashed lines indicating the lifted wires. Figure 5.7d shows the corresponding FEOL graph that achieves security level $k = 1$. Even if all the nets are lifted, the maximum security attainable remains at $k = 1$ as shown in Fig. 5.7e. Figure 5.7f shows the addition of OBISA cells G6, G7, and G8 that are connected to the original design as shown. With the lifting of selected wires, 2-security can be achieved as shown in Fig. 5.7g. An alternative way to attain $k = 2$ security level is to lift all the nets as shown in Fig. 5.7h.

The OBISA wire lifting methodology is shown in Algorithm 35. Steps 1–3 are the same as in Algorithm 34. Then, cells of rarely used types are randomly inserted to compensate for their sparse use and enhance their level of k-security. The cells are placed at random locations. Then, more randomly selected cells are placed at random locations to completely fill the white space. After this, BISA cell routing is performed. Following wire lifting as explained below, backend design is completed.

Algorithm 34: OBISA obfuscated connection methodology [25, 27]

Input: Netlist N
Output: Secure Layout L
   `// Normal physical design`
1 Perform floor-planning, placement and clock-tree synthesis;
   `// Basic BISA insertion`
2 Identify unused space in the layout;
3 Place OBISA cells in the unused space;
4 Connect OBISA cells to form OBISA blocks;
   `// Obfuscation steps`
5 Perform pre-processing;
6 Create fanout nets;
7 Create obfuscation connections;
8 Life wires;
   `// Back to complete normal physical design`
9 Perform routing;
10 Perform signoff functions and generate GDSII;
11 **return** L;

Algorithm 35: OBISA wire lifting methodology [26, 28]

Input: Netlist N
Output: Secure Layout L
   `// Normal physical design`
1 Perform floor-planning, placement and clock-tree synthesis;
   `// Basic BISA insertion`
2 Identify unused space in the layout;
3 Place BISA cells in the unused space;
   `// OBISA steps`
4 Perform compensation cell placement for rare type cells;
5 Perform randomized BISA cell placement;
6 Perform BISA cell routing;
7 Choose wires to lift;
   `// Complete physical design`
8 Route FEOL wires and BEOL wires separately;
9 Perform signoff functions and generate GDSII;
10 **return** L;

Wire lifting is posed as an optimization problem using a binary programming formulation. The goal of optimization is to lift as few wires as possible while achieving k-security for a given k value. Recall that k-security requires that each vertex gate in the FEOL graph is indistinguishable from at least $k-1$ other vertices. If k gates of the same type are indistinguishable from each other, then those k gates satisfy k-security. An assumption is made to simplify the problem: Each vertex will retain at most one edge in the FEOL graph. All other edges will be lifted to BEOL. In a graph that satisfies this property, each vertex has either no edges or a source (sink) with a single edge connecting it to a sink (source). This implies that edges are *uniquely* characterized by the type of the driver gate, the type of the sink gate, and the direction of the edge. This unique characterization constitutes the *type* of an edge. This property is used in the binary programming formulation.

Let k denote the desired k-security level. Let $G = (V, E)$ denote the graph. Let T_v denote the set of cell types. Let T_e denote the set of edge types. Let T_e be partitioned into T_{ea} and T_{ed} that denote the edge types to be allowed and eliminated, respectively. T_{ed} can be determined by counting the number of edges of each type in the graph and determining which types of edges have fewer than k instances. Let $x_i \in \{0, 1\}$, $1 \le i \le |E|$, be the binary valued variables such that $x_i = 0$ denotes lifting of the edge i and $x_i = 1$ denotes the presence of the edge i in FEOL. Let $\mathbf{x} = (x_1, x_2, \ldots, x_{|E|})^T$. The constraints are as follows:

1. Each distinct type of edge has either at least k indistinguishable instances or no instances at all (i.e. completely lifted).

$$\mathbf{A}_{|T_{ea}| \times |E|} \cdot \mathbf{x} \ge k, \quad a_{i,j} = \begin{cases} 1 & \text{edge } j \text{ is of type } i \\ 0 & \text{otherwise} \end{cases} \tag{5.9}$$

$$\mathbf{B}_{|T_{ed}| \times |E|} \cdot \mathbf{x} \le 0, \quad b_{i,j} = \begin{cases} 1 & \text{edge } j \text{ is of type } i \\ 0 & \text{otherwise} \end{cases} \tag{5.10}$$

2. At most one edge can remain for each vertex.

$$\mathbf{C}_{|V| \times |E|} \cdot \mathbf{x} \le 1, \quad c_{i,j} = \begin{cases} 1 & \text{edge } j \text{ is connected to vertex } i \\ 0 & \text{otherwise} \end{cases} \tag{5.11}$$

3. Vertex types must be constrained to have at least k isolated vertices (with all edges lifted).

$$|T_v| - \mathbf{D}_{|V_r| \times |E|} \cdot \mathbf{x} \ge k \tag{5.12}$$

where $d_{i,j}$ is the number of vertices of reference i that edge e_j is connected to.

The optimization goal is to minimize the number of lifted wires or, equivalently, maximize the number of wires left in FEOL:

$$Maximize \sum_{i=1}^{|E|} x_i \tag{5.13}$$

Constraints 5.9, 5.10, and 5.12 imply that at least one solution exists such that all edge types have at least k indistinguishable instances and each gate type has at least k vertices with all edges lifted. However, in practice, all instances of some edge types that have more than k instances in the original graph may have to be lifted to ensure that all others have at least k instances, or all instances of some gate types may keep an edge, thereby leaving no need to lift all edges of at least k instances of each of these types. To accommodate various possible situations, constraints 5.9 and 5.12 are converted into *either-or* constraints [30] using additional binary variables $y_i \in \{0, 1\}, i \in [1..|T_{ea}|]$, and $z_i \in \{0, 1\}, 1 \leq i \leq |V_r|$, to choose between alternatives as follows:

$$\mathbf{A}.\mathbf{x} \geq k - M\mathbf{y} \tag{5.14}$$

$$\mathbf{A}.\mathbf{x} \leq M(\mathbf{y} - 1) \tag{5.15}$$

$$|T_v| - \mathbf{D} \cdot \mathbf{x} \geq k + M(\mathbf{z} - 1) \tag{5.16}$$

$$|T_v| - \mathbf{D} \cdot \mathbf{x} \leq k + M\mathbf{z} \tag{5.17}$$

where $\mathbf{y} = (y_1, y_2, \cdots, y_{|T_{ea}|})^T$, $\mathbf{z} = (z_1, z_2, \cdots, z_{|V_r|})^T$, and M is a large positive number. In the graph model, edges are defined between cells rather than between specific *pins* of the cells. If desired, the binary programming formulation can be modified to accommodate pin based edges without a significant performance penalty.

Algorithm 35 can handle designs consisting of up to several thousand gates. To handle larger designs, hierarchy based partitioning where the design is partitioned according to the constituent module boundaries is performed and the method is applied to each partition separately. For designs that are large but lack hierarchy, geometric partitioning, where the design is partitioned into a suitable number of rectangular regions based on the layout geometry, can be used. Shi et al. [28] provide the details.

5.6.3   Discussion

Xiao et al. [25] have evaluated the OBISA obfuscated connection approach using several circuits from opencores.org. The total number of cells in these benchmarks ranged from 1559 to 49,517 of which 10% to 4% were OBISA cells. OBISA cells are filler cells with no additional area overhead. With M3 as the split layer, 7.6% to 2.7% of the nets were secured by routing through M4 or above. Fanout nets added ranged from 30 to 106. To add about 5% of the obfuscation connections, the number of inputs of the OBISA blocks connected to the original circuit ranged from 15 to 189. This indicates that a relatively small percent of obfuscation connections are sufficient to improve security. Authentication test coverage reaches close to 100% when obfuscation connections are not used. However, as the number of obfuscation connections increases, coverage drops when ATPG vectors are used but can be significantly improved by using several iterations of random patterns where the status of the original circuit (i.e. the values on the obfuscation connections) was changed in each iteration.

Shi et al. [28] have conducted detailed experimental studies to evaluate the OBISA wire lifting method using the selected ISCAS-85 benchmarks and DES and AES cores from opencores.org. They have compared the BP based wire lifting with the greedy wire lifting [1] discussed in Sect. 5.1. For k values ranging from 46 to 4, the OBISA method retained up to 185% of the nets compared to the greedy approach in some cases but lifted more nets than the greedy approach in other cases. OBISA approach is much faster, taking only a few seconds, than the greedy method that often takes several days. In addition, the solution time for the BP formulation is relatively insensitive to the k value.

The pin based method for defining edges proved to be superior to the cell-centered method that resulted in a lower level of security than requested. Experiments showed that the OBISA methods offer competitive protection against the proximity and network flow attacks. In addition, the number of open pins in the FEOL layers can be increased by adding OBISA cells followed by wire lifting. Neighborhood connectivity (see Sect. 3.1) for a given radius is lower with the OBISA insertion and wire lifting when compared with other methods. This indicates excellent spread of the circuit. In addition, standard cell composition bias (Sect. 3.1) could be decreased by 50% after OBISA insertion. Using OBISA insertion and wire lifting on large designs, the worst-case path delay overhead was under 5% and power overhead was under 13%. In addition, high level of security with $k = 160$ and $k = 208$ could be achieved with no additional area required.

The OBISA method is effective in protecting a design not only from trojan insertion attacks but also from cell removal attacks. The OBISA approach demonstrated tight integration of testable structures to improve SM security.

5.7   Summary

In this chapter, we reviewed several design methods to thwart both satisfiability based and design constraint based attacks. In addition to wire lifting and cell insertion, some of these methods combined SM with other DfT (Design for Trust) and DFT (Design for Testability) techniques to significantly enhance security in protecting the ICs from a multiplicity of attack models. Table 5.2 summarizes the defense methods discussed in this chapter.

Table 5.2 Summary of defense methods discussed in Chap. 5

Sl.	Defense method	Year	Attacks thwarted	Benchmarks	Metrics
1	SAT based greedy wire lifting	2013	TI	ISCAS-85, DES	k-Security
2	Simultaneous wire lifting and cell insertion	2018	TI, proximity	ISCAS-85, Functional circuits	k-Security
3	Combined layout camouflaging and SM	2017	TI, SAT, NF, CP	ISCAS-85, ITC-99, EPFL	CCR, HD, OER, E[LS], FOM
4	Combined logic encryption and SM	2019	NF	ISCAS-85, ITC-99	CCR, HD, OER, PNR
5	Combined layout camouflaging and SM for 3D ICs	2018	SAT, 3D proximity	ISCAS-85, ITC-99, DARPA CEP, JPEG	PNR, HD, CCR, k-security
6	Obfuscated Built-In Self-Authentication	2015, 2017	TI, proximity, NF, removal	ISCAS-85, opencores	k-Security, NY, CI, ATPG coverage

TI: Trojan Insertion Attack, NF: Network Flow, SAT: Satisfiability, CP: Crouching Proximity, NY: Neighborhood Connectivity, CI: Composition Index

References

1. F. Imeson, A. Emtenan, S. Garg, M.V. Tripunitara, Securing computer hardware using 3D integrated circuit (IC) technology and split manufacturing for obfuscation, in *Proceedings of the 22nd USENIX Security Symposium* (USENIX Association, New York, 2013), pp. 495–510
2. M. Li, B. Yu, Y. Lin, X. Xu, W. Li, D. Z. Pan, A practical split manufacturing framework for trojan prevention via simultaneous wire lifting and cell insertion, in *Proceedings of the Asia and South Pacific Design Automation Conference, ASP-DAC*, vol. 2018-January (2018), pp. 265–270
3. M. Li, B. Yu, Y. Lin, X. Xu, W. Li, D. Z. Pan, A practical split manufacturing framework for trojan prevention via simultaneous wire lifting and cell insertion. IEEE Trans. Comput. Aided Des. Integr. Circuits Syst. **38**(9), 1585–1598 (2019)
4. J. Cheng, A.W.C. Fu, J. Liu, K-isomorphism: privacy preserving network publication against structural attacks, in *Proceedings of the ACM SIGMOD International Conference on Management of Data* (2010), pp. 459–470
5. J.P. Baukus, L.W. Chow, R.P. Cocchi, B.J. Wang, Method and apparatus for camouflaging a standard cell based integrated circuit with micro circuits and post processing (2012). US Patent, 20120139582
6. S. Patnaik, M. Ashra, J. Knechtel, O. Sinanoglu, Obfuscating the interconnects: low-cost and resilient full-chip layout camouflaging, in *IEEE/ACM International Conference on Computer-Aided Design, Digest of Technical Papers, ICCAD*, vol. 2017-November (Institute of Electrical and Electronics Engineers Inc., New York, 2017), pp. 41–48
7. S. Patnaik, M. Ashraf, O. Sinanoglu, J. Knechtel, Obfuscating the interconnects: low-cost and resilient full-chip layout camouflaging, in *IEEE Transactions on Computer-Aided Design of Integrated Circuits and Systems* (2020), pp. 4466–4481
8. S. Chen, J. Chen, D. Forte, J. Di, M. Tehranipoor, L. Wang, Chip-level anti-reverse engineering using transformable interconnects, in *Proceedings of the 2015 IEEE International Symposium on Defect and Fault Tolerance in VLSI and Nanotechnology Systems* (2015), pp. 109–114

9. M. Tehranipoor, D. Forte, G.S. Rose, S. Bhunia, *Security Opportunities in Nano Devices and Emerging Technologies* (CRC Press, New York, 2017)
10. *Majority-Inverter Graph (MIG)—LSI - EPFL.* https://www.epfl.ch/labs/lsi/page-102566-en-html/mig/
11. J. Magaña, D. Shi, A. Davoodi, Are proximity attacks a threat to the security of split manufacturing of integrated circuits? in *IEEE/ACM International Conference on Computer-Aided Design, Digest of Technical Papers, ICCAD*, vol. 07-10-November (2016), pp. 1–7
12. A. Sengupta, M. Nabeel, J. Knechtel, O. Sinanoglu, A new paradigm in split manufacturing: lock the FEOL, unlock at the BEOL, in *Proceedings of the Design, Automation and Test in Europe Conference and Exhibition* (2019), pp. 414–419
13. A. Sengupta, M. Nabeel, M. Yasin, O. Sinanoglu, ATPG-based cost-effective, secure logic locking, in *Proceedings of the IEEE VLSI Test Symposium*, vol. 2018-April (2018), pp. 1–6
14. Y. Wang, P. Chen, J. Hu, J.J. Rajendran, The cat and mouse in split manufacturing, in *Proceedings of the Design Automation Conference*, vol. 05-09-June (2016), pp. 1–6
15. Y. Wang, P. Chen, J. Hu, J. Rajendran, Routing perturbation for enhanced security in split manufacturing, in *Proceedings of the Asia and South Pacific Design Automation Conference, ASP-DAC*, vol. 2 (Institute of Electrical and Electronics Engineers Inc., New York, 2017), pp. 605–610
16. S. Patnaik, J. Knechtel, M. Ashraf, O. Sinanoglu, Concerted wire lifting: enabling secure and cost-effective split manufacturing, in *Proceedings of the Asia and South Pacific Design Automation Conference* (2018), pp. 251–258
17. S. Patnaik, M. Ashraf, J. Knechtel, O. Sinanoglu, Raise your game for split manufacturing: restoring the true functionality through BEOL, in *Proceedings of the55th ACM/ESDA/IEEE Design Automation Conference*, vol. 6 (IEEE, New York, 2018), pp. 1–6
18. S. Patnaik, M. Ashraf, O. Sinanoglu, J. Knechtel, Best of both worlds: integration of split manufacturing and camouflaging into a security-driven CAD flow for 3D ICs, in *Proceedings of the IEEE/ACM International Conference on Computer-Aided Design, Digest of Technical Papers*, vol. 11 (2018), pp. 1–8
19. S. Patnaik, M. Ashraf, O. Sinanoglu, J. Knechtel, A Modern approach to IP protection and trojan prevention: split manufacturing for 3D ICs and obfuscation of vertical interconnects, in *IEEE Transactions on Emerging Topics in Computing* (2019), pp. 1–18
20. Common evaluation platform, in *Assistant Secretary of Defense for Research and Engineering* (2019). https://github.com/mit-ll/CEP
21. Opencores, in *Reference Community for Free and Open Source IP cores* (2020). https://opencores.org/
22. S. Patnaik, *GitHub - DfX-NYUAD/3D-SM-Attack: Proximity Attack for 3D ICs with Obfuscated F2F Mappings.* https://github.com/DfX-NYUAD/3D-SM-Attack
23. K. Xiao, M. Tehranipoor, BISA: Built-in self-authentication for preventing hardware trojan insertion, in *Proceedings of the IEEE International Symposium on Hardware-Oriented Security and Trust* (2013), pp. 45–50
24. K. Xiao, D. Forte, M. Tehranipoor, A novel built-in self-authentication technique to prevent inserting hardware trojans. IEEE Trans. Comput. Aided Des. Integr. Circuits Syst. **33**(12), 1778–1791 (2014)
25. K. Xiao, D. Forte, M.M. Tehranipoor, Efficient and secure split manufacturing via obfuscated built-in self-authentication, in *Proceedings of the IEEE International Symposium on Hardware-Oriented Security and Trust* (2015), pp. 14–19
26. Q. Shi, K. Xiao, D. Forte, M.M. Tehranipoor, Securing split manufactured ICs with wire lifting obfuscated built-in self-authentication, in *Proceedings of the ACM Great Lakes Symposium on VLSI, GLSVLSI*, vol. Part F1277 (2017), pp. 339–344
27. Q. Shi, K. Xiao, D. Forte, M.M. Tehranipoor, Obfuscated built-in self-authentication, in *Hardware Protection through Obfuscation* (Springer, Berlin, 2017), pp. 263–289

28. Q. Shi, M.M. Tehranipoor, D. Forte, Obfuscated built-in self-authentication with secure and efficient wire-lifting. IEEE Trans. Comput. Aided Des. Integr. Circuits Syst. **38**(11), 1981–1994 (2019)
29. M. Bushnell, V.D. Agrawal, *Essentials of Electronic Testing for Digital, Memory and Mixed-Signal VLSI Circuits* (Springer, Berlin, 2002)
30. M.-K. Kim, B.A. McCarl, T.H. Spreen, Applied mathematical programming, in *Textbooks*, vol. 6 (2018). https://digitalcommons.usu.edu/oer_textbooks/6/

Chapter 6
Challenges and Research Directions

Abstract In this chapter, we discuss several challenges and research directions. Specifically in the context of the methods discussed in this book, we review the need for SM benchmarks, trojan insertion and detection experiments, attack validation methods, and practical demonstrations. We also discuss several emerging and new research directions, including SM methods at higher levels of abstractions, SM for analog and mixed-signal designs, use of novel devices in SM, new attack models against SM, simultaneous optimization of SM manufacturability and security, and application of advances in ML for SM attacks and defenses.

6.1 Challenges

In this book, we have discussed a wide range of attack methods against SM and defense techniques developed in response, representing progress made over the past decade. A recent and timely survey by Perez et al. [1] provides a succinct summary of and comparison among some of these attacks and defense methods. Both attack and defense methods have been developed for both reverse engineering and *trojan insertion* attack models in the context of 2D, 2.5D, and 3D designs. However, some challenges remain. Foremost among them that could provide further impetus to SM research along its current trajectory are the following:

1. *Benchmarks:* As evidenced by the discussions in this book, most, but not all, SM attacks and defenses are evaluated using benchmark sets that have one or more of the following limitations: (1) relatively small size circuits, (2) exclusively combinational logic netlists, and (3) benchmarks designed for some other purposes (e.g. logic synthesis, testing, etc.) many years ago.

   Development of realistic, large size benchmarks specifically to challenge SM attacks and defense methods for 2D, 2.5D, and 3D ICs would be of immense value for uniform evaluation and comparison of SM techniques.

2. *Trojan Insertions:* Trojan insertion at the FEOL foundry is a widely used attack model and several attacks proposed netlist to layout mapping methods for trojan insertion attacks. However, exactly what type of nodes could be targeted by

© The Author(s), under exclusive license to Springer Nature Switzerland AG 2021
R. Vemuri, S. Chen, *Split Manufacturing of Integrated Circuits for Hardware Security and Trust*, https://doi.org/10.1007/978-3-030-73445-9_6

attackers, the type of trojans that can be inserted at the successfully mapped nodes, how they may be able to avoid detection from the existing trojan detection methods, how effective the proposed SM defense methods are in mitigating trojan insertion attacks, and what new defense methods might be needed are all open questions. Attempting to clearly answer these questions could open new research directions.

3. *Metrics vs. Validation Methods:* Many metrics were proposed to evaluate the effectiveness of various SM attacks and defenses. While these are all useful metrics to evaluate and compare the attacks and defenses from the researchers' viewpoint, in general, they cannot be used by the attackers to determine whether the attack is successful. In the absence of an oracle, methods the *attackers* can use to quickly and inexpensively determine the (degree of) correctness of the recovered design would be valuable in developing new insights into both attacks and defenses.

4. *Practical Demonstrations:* After some initial success stories, discussed in Chap. 1, there have been no further practical demonstrations of SM. Demonstrations and studies of 2D, 2.5D, and 3D SM techniques mixing multiple fabrication processes would be invaluable to understand the cost, PPA, reliability, and yield tradeoffs, to further evaluate foundry compatibility issues, and to establish the overall manufacturability guidelines. In addition, practical demonstrations of attacks and defense methods are necessary to justify potential deployment of SM for security and trust.

In the next few sections, we discuss several new and emerging directions.

6.2 Splitting at Higher Levels of Abstraction

Most of the SM methods proposed to date operate at the logic level or layout level of abstraction in the design process. Some of these methods require expensive re-synthesis cycles. SM methods at higher levels of abstraction are worth exploring to reduce time while potentially enhancing security-cost tradeoff exploration. Effective methods for high level signal lifting are yet to be developed. Higher levels of abstraction include register-transfer level, algorithmic behavior level, and system-on-chip specification level.

Cui et al. [2] recently proposed a register-transfer level (RTL) obfuscation method based on SM. In this method, connections among functional units are lifted into the hidden layers. An attacker with only the FEOL information would see a sea of functional units. In addition, dummy multiplexors at the inputs of selected functional units are added. These MUXes serve the purpose of obfuscating the real connections by increasing the number of data paths that can be recovered and need not be actually connected. To further increase the number of recoverable data paths, dummy functional units can be added if the area budget allows them. The method is primarily aimed at 2.5D and 3D ICs but can also be adapted to 2D ICs. Cui et al. theoretically proved the high security and low-cost benefits and experimentally demonstrated split designs using the 2.5D stacked silicon interconnect (SSI) based Xilinx FGPA devices [3]. Cui et al.'s work is a first step toward exploring the benefits and drawbacks of splitting at higher abstraction levels.

Several researchers recognized the benefits of 3D integration for developing secure architectures. Sect. 1.11 mentions some of the early architectures proposed. Robust splitting methods to identify the signals to be hidden in the RDLs are needed at the level of system design along with appropriate evaluation metrics.

Similarly, attacks that can work at higher levels of inference are needed especially in the context of 3D integration in view of the possibility that all foundries involved may have to be assumed to be untrusted. Conversely, how to defend a design in such a scenario while containing the cost and PPA overhead is a fruitful direction for further research.

6.3   SM for Analog and Mixed-Signal Designs

This book exclusively focuses on SM for digital designs. In modern electronic systems, some analog modules are unavoidable, especially in the front-ends and back-ends of communication, embedded, and cyber-physical systems. SM methods can be beneficial in the context of analog, RF, and mixed-signal designs too.

Analog and *RF designs* [4–8] consist of passive components such as resistors, capacitors, and inductors. These passive components are often implemented fully or partially in metal layers. For example, capacitors are often implemented in upper-level metal layers and inductors in the top metal layer. Wire lengths and directions play a key role in determining the function and performance parameters of the components implemented. If the upper metal layers are delegated to the BEOL foundry, then attackers having access only to the FEOL layout are confronted with missing components the recovery of which may require searching through an extremely large, potentially infinite, solution space. Unlike in the case of digital designs, proximity based attacks are unlikely to succeed for RF design recovery. Due to these reasons, SM can be used quite effectively to secure analog and RF designs.

Based on these observations, Bi et al. [9, 10] proposed SM for RF designs. They have considered three SM scenarios: (1) Remove the top metal layer that removes all inductors. (2) Remove the top two metal layers that would remove both inductors and capacitors. (3) Add additional obfuscation by enlarging the area of the passive components and by adding empty blocks in the design. These measures make it hard for the attackers to postulate the locations and values of the inductors and capacitors. They have applied these methods to a class AB power amplifier and a class E power amplifier implemented in a 180 nm process with 6 metal layers. Considering the output power and power-added efficiency of the power amplifiers as the characteristics being hidden from the attacker, they have reported extensive simulation studies to support the efficacy of SM for RF design security.

Further research in this area should consider analog and mixed-signal designs since analog subsystems are present in most electronics for communication applications and for embedded systems with analog sensor or actuator interfaces.

6.4 SM with Novel Devices

As the MOS transistor scaling is reaching its limits, numerous CMOS-compatible and beyond-CMOS devices are being proposed and evaluated for use in future ICs. Some of these devices have performance, power, and area benefits, while others offer security benefits due to their polymorphic or stochastic behavior. The following are examples of SM applications of some novel devices:

1. *Resistive RAMs:* A resistive random-access memory (RRAM or ReRAM) device is a non-volatile memory element whose terminal resistance across a solid-state dielectric material can be changed by the application of an electric pulse [11, 12]. RRAM devices offer excellent area and power advantages and compatibility for monolithic integration with CMOS.

   Liauw et al. [13] proposed a switch-box design using RRAMs as part of a 3D FPGA architecture that uses an RRAM-based configuration memory. Wu et al. [14, 15] suggested using the RRAM-based switch boxes as obfuscation elements to introduce incorrect connections at several randomly selected locations in the design to protect the trojan detection circuitry. RRAM cells can be integrated into the BEOL metal layers, making this method area-efficient and compatible with SM. When using SM, while switch-box programmability is not necessary, the correct configuration is obfuscated from a prospective attacker at the FEOL foundry. Wu et al. demonstrated this method by designing and fabricating a 65 nm LZ encoder chip [14].

2. *Floating-Gate MOS Transistors:* Floating-gate MOS (FGMOS) transistors [16] are MOSFETs in which the gate is electrically isolated. Electrons placed on this "floating" gate are trapped resulting in the non-volatile state of the switch. A secondary control gate, electrically isolated from the floating gate, but capacitively connected to it, is used for programming. FGMOS transistors have been used in non-volatile memories such as EPROMs, EEPROMs, and flash memories and more recently, in neural networks, analog and mixed-signal circuits, and tunable and reconfigurable circuits [17, 18].

   Madani et al. [19] proposed using FGMOS devices to obfuscate selected nets in split manufactured 3D ICs. In their method, an extra security tier is introduced between neighboring functional tiers in a 3D structure. Critical nets are selected and, for each critical net, the source and sink gates are placed in different functional layers. An OR-gate obfuscator is introduced to mix the genuine driver with other randomly selected "bogus" drivers as shown in Fig. 6.1a. FGMOS transistors are introduced on each input of the OR gate. All the FGMOS transistors are lifted into the security layer as shown in Fig. 6.1b and

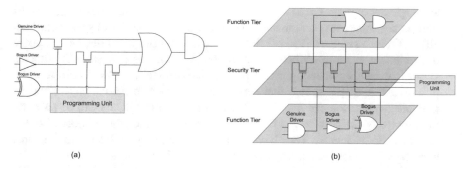

Fig. 6.1 Security tier using FGMOS transistors in a 3D IC (based on [19]). (**a**) Mixing of drivers, (**b**) Lifting FGMOS transistors

are programmed using an external programming unit after the 3D IC is produced. In this way, security is guaranteed regardless of where the functional and security tiers are fabricated and integrated. All the foundries may be untrusted. Since no foundry knows the correct connections for programming, the design cannot be attacked either for reverse engineering or for trojan insertion if critical nets are selected and obfuscated.

3. *Vertical Slit Field-Effect Transistors:* Vertical Slit Field-Effect Transistors (VeS-FET) [20, 21] are twin-gate transistors consisting of a horizontal channel and four metal pillars implementing vertical terminals (two gates and two drain/source terminals). Transistors are fabricated as regular array structures called *canvases*. VeSFET terminals can be accessed from both ends of the pillars that enable two-sided routing and make them highly suitable for monolithic 3D integration. In place of the TSVs, the pillars naturally provide vertical wiring channels. Taking advantage of their two-sided routability and 3D friendliness, Yang et al. [22] proposed SM approaches for VeSFET based 2D and 3D ICs. They proposed a two-way design partitioning method such that the design can be implemented by two foundries both of which may be untrusted. All devices and the majority of connections would be manufactured at one foundry and a selected few connections would be realized at the other foundry. In addition, for 3D circuits, some of the transistors can be implemented in the lower tiers hidden from the first foundry. The second foundry uses these transistors to implement a portion of the circuit. Thus, neither foundry has access to the complete design. In 2D VeSFET ICs, trojan detection can be done by measuring the leakage current that shows significant change if any of the existing transistors are moved or new transistors are added. No additional circuitry is needed for trojan detection. In 3D VeSFET ICs, a scan path implementation is proposed to facilitate trojan detection. Finally, the overall security is enhanced by a simple logic encryption method. Yang et al. evaluated these methods using circuits from the MCNC benchmark suite.

4. *Giant Spin-Hall Effect Switches:* Several novel devices can be used to implement polymorphic gates that can implement different Boolean functions based on the control input during run-time. Polymorphic gates readily support camouflaging

and logic encryption. Polymorphism supports the former and the control inputs used as key inputs support the latter. Utilizing these properties of giant spin-Hall effect (GSHE) switches [23, 24], Patnaik et al. [25] proposed designing secure deterministic and probabilistic circuits that can defend against SAT attacks, various reverse engineering attacks, and side-channel attacks. Their security primitive is a GSHE switch that can be programmed to realize any of the 16 possible 2-input Boolean functions. Wires for the control inputs will be partially or completely assigned to the BEOL layers to defeat the attackers based at the untrusted FEOL facility. They have analyzed the security of this method against a variety of attacks using a range of small to large *benchmarks*.

Numerous other technologies ranging from carbon nanotube (CNT) interconnects and CNT field-effect transistors (CNTFETs) to spin-based storage and computing devices are on the verge of leaving research labs into commercial products. These technologies offer interesting opportunities and new challenges for security including SM.

6.5   Novel Attacks Against SM

Most of the attack models discussed in this book assumed that the BEOL foundry is safe and the FEOL foundry is untrusted. Wang et al. [26] considered potential attacks at an untrusted BEOL foundry to reconstruct the FEOL netlist. The attacker is assumed to know the cell library used in the design. Transistors and the M1 metal layer are assumed to be in the FEOL and the rest (including M1–M2 vias) in the BEOL that provides maximum protection from an attacker at the FEOL foundry. They have proposed a *geometric pattern matching* method that examines the M1–M2 via locations in BEOL and matches them with the M1 patterns of the standard cells in the library. The basic idea is illustrated in Fig. 6.2. A matching

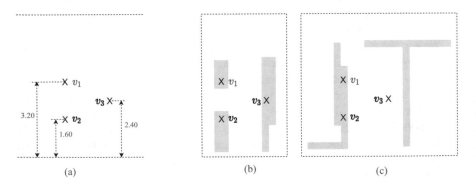

Fig. 6.2 Geometric pattern matching. (**a**) M1–M2 via locations in BEOL (**b**) A cell with matching M1 patterns (**c**) A mismatched cell (based on [26])

procedure constructs a set of plausible matches considering a row of cells at once. The procedure considers geometric shifting via patterns to find cell matches while determining cell boundaries to maximize the matches. Since some M1–M2 vias may match multiple cells, a pruning method is used to prune the matches based on the layout context. Experimental results using the ISCAS-85 and ITC-99 benchmarks show that 100% of the FEOL cells are correctly matched, indicating that this attack is highly effective. They have also proposed another attack based on machine learning that was also shown to be almost as effective.

Wang et al. proposed a defense method in which all the cells are modified to have identical M1 patterns. This will ensure that the attacker cannot differentiate between the cells based on the M1–M2 via footprints in BEOL. However, this method has excessive performance penalty, 31.66% on average. Hence, two technology mapping strategies are proposed to meet performance vs. security priority.

This attack method assumed that split is done above M1 and exploited the M1–M2 via footprints. It is unclear if the attack would be similarly effective if splitting is done above several metal layers, say at M5.

Novel attacks against SM must be explored in the context of (1) 3D integration, (2) splitting at higher levels, and (3) use of novel devices in SM.

6.6   SM Manufacturability vs. Security

Some defense methods discussed in this book considered PPA costs while making security-driven design decisions. BEOL manufacturing costs are considered indirectly in terms of the split level or the number of BEOL nets. Manufacturability is an important consideration while designing an IC, especially during the layout synthesis stage. Among the manufacturability issues are CMP (Chemical-Mechanical Planarization) uniformity and SADP (Self-Aligned Double Patterning) compliance. Feng et al. [27] proposed methods for improving SM security as well as CMP uniformity or SADP compliance as follows: Both lifting of nets from FEOL to BEOL and rerouting of wires in FEOL affect wire density as well as SM security. Variations in wire density impact CMP. Based on this observation, Feng et al. proposed a method for selecting nets to be lifted and a wire rerouting method to improve both SM security and CMP uniformity.

The SADP process first prints a sea of parallel wires in a routing plane and then uses a second mask to cut the wires into wire segments that are used to carry various signals. SADP compliance requires honoring a minimum spacing rule between the cuts. Ends of some wire segments are extended after the initial routing to honor this minimum spacing [27, 28]. Extension of a wire segment could impact security either favorably or unfavorably depending on the direction and context of extension. Feng et al. proposed a method based on net lifting followed by FEOL wire extension for simultaneous improvement of SM security and SADP minimum cut spacing compliance. Wire extension is formulated as an ILP (Integer Linear Programming) problem.

Experimenting with the ISCAS-85 and ITC-99 benchmarks, Feng et al. demonstrated that their CMP-security method, while simultaneously improving security, can reduce wire density variance by 37% for FEOL and 25% for BEOL, whereas the traditional security-only methods would increase it by 18% for BEOL. Similarly, their SADP-security method can reduce cut rule violations by 97%, while the traditional security-only methods would increase the violations by 44%.

Recall that SM was originally proposed as a means for enhancing manufacturing yield. Further research is warranted to develop cost models of yield, reliability, and manufacturability in the context of SM for 2D, 2.5D, and 3D ICs and simultaneous optimization of those costs along with the traditional security and PPA metrics.

6.7   Advances in Machine Learning for SM

Numerous machine learning methods are being applied to complex, large-scale engineering design problems. In particular, advances in *deep learning* (DL) [29, 30] already revolutionized many application areas. When DL is applied to a problem, appropriate feature extraction, neural network architecture definition, sample selection, training, and validation are all among the challenges to be tackled before successful use of the network for new problem instances.

Li et al. [31, 32] recently proposed a DL architecture to attack split IC designs at the FEOL foundry. They have defined a fairly complex *deep neural network* (DNN) to simultaneously process both vector-based and image-based layout features. Vector-based features include various distances related to virtual pin pairs, load capacitances and number of sink nodes, FEOL wire lengths and vias, and driver delays. Image-based features include, for each virtual pin, the grey-scale images of the FEOL layout region centered around the pin. Three different scales with the same image shape at three different precisions are considered. The output data consists of scores for every pair of virtual pins. Given a set of virtual pin pairs, the network predicts the connection probability for each pair. A softmax regression loss is proposed such that the most probable BEOL connection can be directly selected. Using the ISCAS-85, MCNC, and ITC-99 benchmark suites, Li et al. demonstrated $1.21\times$ and $1.12\times$ improvement in CCR over the network flow attack, discussed in Sect. 2.5, when the splitting is done at M1 and M3, respectively, taking less than 1% of its execution time. When compared with the ML attack discussed in Sect. 2.6, Li's DL method reduced the potential candidate list by 47% with only 1% loss of accuracy while improving CCR by $2.2\times$. Li et al. also proposed a random routing blockage insertion method, similar to the one discussed in Sect. 3.11, to defeat their DL based attack while incurring minimal power and performance overheads. They have made the source codes for feature extraction and DNN and the protected layouts available at [33–35].

DL attacks on split designs need to be further improved to successfully attack designs protected with routing perturbations and demonstrated on industrial scale problem instances. Robust DL methods should be relatively insensitive to noise in the training data and should provide explanation for their decisions. Using such methods for SM attacks and defenses could provide a means to tackle large-scale designs. Representing progress in this direction, Zeng et al. [36] recently reported a novel routing obfuscator for SM based on explanations from the ML attack discussed in Sect. 2.6.

6.8 Summary

Research in split manufacturing attacks and defense methods has progressed rapidly in the past decade, with several new methods appearing just in the past 2 years. We have presented a summary of the state-of-the-art methods in this area in the previous chapters and some challenges and emerging directions in this chapter. Effectiveness and viability of SM as a part of mitigating various IC supply chain risks will be determined in the coming years based both on the research results and on the manufacturing trends and economics.

References

1. T.D. Perez, S. Pagliarini, A survey on split manufacturing: attacks, defenses, and challenges. IEEE Access **8**(10), 184013–184035 (2020)
2. X. Cui, J.J. Zhang, K. Wu, S. Garg, R. Karri, Split manufacturing-based register transfer-level obfuscation. ACM J. Emerging Technol. Comput. Syst. **15**(1), 1–22 (2019)
3. L. Madden, S. Ramalingam, X. Wu, E. Wu, B. Banijamali, N. Kim, K. Abugharbieh, Xilinx stacked silicon interconnect technology delivers breakthrough FPGA performance. Adv. Microelectron. **40**(3), 6–11 (2013)
4. B. Razavi, *Design of Analog CMOS Integrated Circuit*, 2nd edn. (McGraw Hill, New York, 2017)
5. Johns, D.A., Ken M., *Analog Integrated Circuits Design*, 2nd edn. (Wiley, London, 2012)
6. R.J. Baker, *CMOS: Circuit Design, Layout, and Simulation*, 4th edn. (Wiley, London, 2019)
7. T. H. Lee, *The Design of CMOS Radio-Frequency Integrated Circuits*, vol. 12 (Cambridge University Press, Cambridge, 2003).
8. A. Hastings, *The Art Of Analog Layout*, 2nd edn. (Pearson/Prentice Hall, Upper Saddle River, 2006)
9. Y. Bi, S. Member, J.-S. Yuan, S. Member, Y. Jin, Split manufacturing in radio-frequency designs, in *International Conference on Security and Management* (2015), pp. 204–210
10. Y. Bi, J.S. Yuan, Y. Jin, Beyond the interconnections: split manufacturing in RF designs. Electronics **4**(3), 541–564 (2015)
11. F. Zahoor, T.Z. Azni Zulkifli, F.A. Khanday, Resistive random access memory (RRAM): an overview of materials, switching mechanism, performance, multilevel cell (MLC) storage, modeling, and applications. Nanoscale Res. Lett. **15**, 90 (2020)
12. H.Y. Chen, S. Brivio, C.C. Chang, J. Frascaroli, T.H. Hou, B. Hudec, M. Liu, H. Lv, G. Molas, J. Sohn, S. Spiga, V.M. Teja, E. Vianello, H.S. Wong, Resistive random access memory (RRAM) technology: from material, device, selector, 3D integration to bottom-up fabrication. J. Electroceram. **39**, 1–4 (2017)

13. Y.Y. Liauw, Z. Zhang, W. Kim, A.E. Gamal, S.S. Wong, Nonvolatile 3D-FPGA with monolithically stacked RRAM-based configuration memory, in *Digest of Technical Papers—IEEE International Solid-State Circuits Conference*, vol. 55 (2012), pp. 406–408

14. T.F. Wu, K. Ganesan, Y.A. Hu, H.S. Wong, S. Wong, S. Mitra, TPAD: hardware trojan prevention and detection for trusted integrated circuits. IEEE Trans. Comput. Aided Des. Integr. Circuits Syst. **35**(4), 521–534 (2016)

15. S. Mitra, H.S. Wong, S. Wong, The Trojan-proof chip (2015), pp. 46–51

16. P. Hasler, B.A. Minch, C. Diorio, Floating-gate devices: they are not just for digital memories anymore, *Proceedings—IEEE International Symposium on Circuits and Systems*, vol. 2 (1999), pp. 388–391

17. E. Ozalevli, P. Hasler, *Floating-Gate Transistors in Analog and Mixed-Signal Circuit Design: Programming, Design Methodology, and Applications* (VDM Verlag Dr. Müller, 2009)

18. E. Ozalevli, *Tunable and Reconfigurable Circuits Using Floating-Gate Transistors: Programming and Tuning, Design Method, Applications* (VDM Verlag Dr. Müller, 2009)

19. S. Madani, M.R. Madani, I. Kalyan Dutta, Y. Joshi, M. Bayoumi, A hardware obfuscation technique for manufacturing a secure 3D IC, in *Midwest Symposium on Circuits and Systems* (Institute of Electrical and Electronics Engineers, Piscataway, 2019), pp. 318–323

20. W. Maly, N. Singh, Z. Chen, N. Shen, X. Li, A. Pfitzner, D. Kasprowicz, W. Kuzmicz, Y.W. Lin, M. Marek-Sadowska, Twin gate, vertical slit FET (VeSFET) for highly periodic layout and 3D integration, in *Proceedings of the 18th International Conference—Mixed Design of Integrated Circuits and Systems, MIXDES 2011* (2011), pp. 145–150

21. P.L. Yang, T.B. Hook, P.J. Oldiges, B.B. Doris, Vertical slit FET at 7-nm node and beyond. IEEE Trans. Electron Devices **63**(8), 3327–3334 (2016)

22. P.L. Yang, M. Marek-Sadowska, Making split-fabrication more secure, in *IEEE/ACM International Conference on Computer-Aided Design, Digest of Technical Papers, ICCAD*, vol. 07-10-Nove (Institute of Electrical and Electronics Engineers, Piscataway, 2016), pp. 1–6

23. S. Datta, S. Salahuddin, B. Behin-Aein, Non-volatile spin switch for Boolean and non-Boolean logic. Appl. Phys. Lett. **101**(25), 252411–252420 (2012)

24. N. Rangarajan, A. Parthasarathy, N. Kani, S. Rakheja, Energy-efficient computing with probabilistic magnetic bits—performance modeling and comparison against probabilistic CMOS logic. IEEE Trans. Magn. **53**(11), 1–10 (2017)

25. S. Patnaik, N. Rangarajan, J. Knechtel, O. Sinanoglu, S. Rakheja, Advancing hardware security using polymorphic and stochastic spin-hall effect devices, in *Proceedings of the 2018 Design, Automation and Test in Europe Conference and Exhibition, DATE 2018*, vol. 2018-January, 2018, pp. 97–102

26. Y. Wang, T. Cao, J. Hu, J. Rajendran, Front-end-of-line attacks in split manufacturing, in *IEEE/ACM International Conference on Computer-Aided Design, Digest of Technical Papers, ICCAD*, vol. 2017-Novem, 2017, pp. 1–8

27. L. Feng, Y. Wang, J. Hu, W.K. Mak, J. Rajendran, Making split fabrication synergistically secure and manufacturable, in *IEEE/ACM International Conference on Computer-Aided Design, Digest of Technical Papers, ICCAD*, 2017, pp. 313–320

28. Y. Ding, C. Chu, W.K. Mak, Throughput optimization for SADP and e-beam based manufacturing of 1d layout, in *Proceedings—Design Automation Conference* (2014), pp. 1–6

29. S. Skansi, *Introduction to Deep Learning—From Logical Calculus to Artificial Intelligence* (Springer, Berlin, 2018)

30. U. Michelucci, *Applied Deep Learning: A Case-Based Approach to Understanding Deep Neural Networks* (Apress, New York, 2018)

31. H. Li, S. Patnaik, A. Sengupta, H. Yang, J. Knechtel, B. Yu, E.F. Young, O. Sinanoglu, Attacking split manufacturing from a deep learning perspective, in *Proceedings—Design Automation Conference*, vol. 6 (2019), pp. 1–6

32. H. Li, S. Patnaik, M. Ashraf, H. Yang, J. Knechtel, B. Yu, O. Sinanoglu, E.F. Young, Deep learning analysis for split manufactured layouts with routing perturbation, in *IEEE Transactions on Computer-Aided Design of Integrated Circuits and Systems* (2020), pp. 1–14

33. GitHub—cuhk-eda/split-extract: Heterogeneous Feature Extraction for Split Manufactured Layouts with Routing Perturbation (2020). https://github.com/cuhk-eda/split-extract

34. GitHub–cuhk-eda/split-attack: Deep Learning Analysis for Split Manufactured Layouts with Routing Perturbation (2020). https://github.com/cuhk-eda/split-attack

35. S. Patnaik, GitHub—DfX-NYUAD/Randomized_routing_perturbation (2020). https://github.com/DfX-NYUAD/Randomized_routing_perturbation

36. W. Zeng, A. Davoodi, R.O. Topaloglu, ObfusX: routing obfuscation with explanatory analysis of a machine learning attack, in *Proceedings of the 26th Asia and South Pacific Design Automation Conference* (ACM, New York, 2021), pp. 548–554

Appendix A
Benchmarks

Researchers have used various benchmark sets to evaluate and compare the attack methods and the defense methods. In this appendix, we summarize the features of these benchmark suites and cite sources for additional information and for obtaining the benchmarks themselves. These benchmark sets are referenced throughout this book.

1. *ISCAS-85:* This set of 10 combinational logic circuits was made available to the authors at the International Symposium on Circuits and Systems (ISCAS), 1985. These were described by Brglez et al. in [1] and partially characterized in [2]. A small circuit named c17, which was used as an illustrative example in the documentation [3], has also been used by researchers. High-level descriptions of these benchmarks were reverse engineered by Hansen et al. [4] and made available from [5]. Features of these 11 circuits are summarized in Table A.1. These benchmarks are available via [6, 7].

Table A.1 ISCAS-85 benchmark circuits

Circuit	Function	#Inputs	#Outputs	#Gates	#Nets
c17[a]	Logic	5	2	6	11
c432	27-channel interrupt controller	36	7	160	196
c499	32-bit SEC circuit	41	32	202	243
c880	8-bit ALU	60	26	383	443
c1355	32-bit SEC circuit	41	32	546	587
c1908	16-bit SEC/DED circuit	33	25	880	913
c2670	12-bit ALU and controller	157	64	1193	1350
c3540	8-bit ALU	50	22	1669	1719
c5315	9-bit ALU	178	123	2307	2485
c6288	16x16 multiplier	32	32	2416	2448
c7552	32-bit adder/comparator	207	108	3512	3719

[a]Used as an example in the documentation

© The Author(s), under exclusive license to Springer Nature Switzerland AG 2021
R. Vemuri, S. Chen, *Split Manufacturing of Integrated Circuits for Hardware Security and Trust*, https://doi.org/10.1007/978-3-030-73445-9

Table A.2 ISCAS-89 benchmark circuits

Circuit	#Inputs	#Outputs	#Gates	#FFs	#Nets
s27	4	1	10	3	17
s208	11	2	96	8	115
s298	3	6	119	14	136
s344	9	11	160	15	184
s349	9	11	161	15	184
s382	3	6	158	21	182
s386	7	7	159	6	172
s400	3	6	162	21	186
s4240	19	2	196	16	231
s444	3	6	181	21	205
s510	19	7	211	6	236
s526	3	6	193	21	217
s641	35	24	379	19	433
s715	35	23	493	19	547
s820	18	19	289	5	312
s832	18	19	287	5	310
s838	35	2	390	32	457
s953	16	23	395	29	440
s1196	14	14	529	18	561
s1238	14	14	508	18	540
s1423	17	5	657	74	748
s1488	8	19	653	6	667
s1494	8	19	647	6	661
s5378	35	49	2779	179	2993
s9234	36	39	5597	211	5844
s13207	62	152	7951	638	8651
s15850	77	150	9772	534	10,383
s35932	35	320	16,065	1728	17,828
s38417	28	106	22,179	1636	23,843
s38584	38	304	19,253	1426	20,717

2. *ISCAS-89:* This is a set of 31 benchmarks originally distributed to the participants of the Special Session on Sequential Test Generation at the International Symposium on Circuits and Systems (ISCAS), 1989 [8]. These circuits were partially characterized by Brglez et al. in [9] and are available from [6, 7]. Features of these circuits are summarized in Table A.2.

3. *MCNC:* Microelectronics Center of North Carolina (MCNC) used to distribute the ISCAS-85 and ISCAS-89 sets. Additional benchmarks were developed in conjunction with various Logic Synthesis workshops. Several of these were distributed as LGSynth-89, LGSynth-91, and LGSynth-93 bundles. Subsets of these benchmarks are known variously as the *MCNC Benchmarks* or the *ACM/SIGDA Benchmarks*. Table A.3 summarizes the features of 5 combinational

Table A.3 MCNC
benchmark circuits

Circuit	#Inputs	#Outputs	#Gates	#Nets
apex2	39	3	610	649
apex4	10	19	5360	5370
des	256	245	6473	6729
ex1010	10	10	5066	5076
seq	41	35	1459	1472

Table A.4 ITC-99 benchmark circuits

Circuit	Function	#Inputs	#Outputs	#Gates	#FFs	#Nets
b01	FSM that compares serial flows	2	2	49	5	56
b02	FSM that recognizes BCD numbers	1	1	28	4	33
b03	Resource arbiter	4	4	160	30	194
b04	Compute min and max	11	8	737	66	814
b05	Elaborate the contents of a memory	1	36	998	34	1033
b06	Interrupt handler	2	6	56	9	67
b07	Count points on a straight line	1	8	441	49	491
b08	Find inclusions in sequences of numbers	9	4	183	21	213
b09	Serial to serial converter	1	1	170	28	199
b10	Voting system	11	6	172	17	200
b11	Scramble string with variable cipher	7	6	726	31	764
b12	1-player game (guess a sequence)	5	6	944	121	1070
b13	Interface to meteo sensors	10	10	362	53	425
b14	Viper processor (subset)	32	54	10,098	245	10,375
b15	80386 processor (subset)	36	70	8922	449	9407
b17	Three copies of b15	37	97	32,326	1415	33,778
b18	Two copies of b14 and two of b17	37	23	114,621	3320	117,978
b19	Two copies of b14 and two of b17	24	30	231,320	6642	237,986
b20	A copy of b14 and a modified version of b14	32	22	20,226	490	20,748
b21	Two copies of b14	32	22	20,571	490	21,093
b22	A copy of b14 and two modified versions of b14	32	22	29,951	735	30,718

circuits used under the name *"MCNC Benchmarks"* in some SM papers. These benchmarks are described in [10] and can be obtained from [6, 7].

4. *ITC-99:* This is a set of 22 sequential circuits obtained from various companies and released at the International Test Conference (ITC), 1999 [11–13]. These circuits were discussed by Corno et al. in [14] and can be obtained from [6, 12]. Features of 21 of these benchmarks used in SM research are summarized in Table A.4.

5. *ISPD-11:* This set consists of 8 benchmarks originally developed for a routability-driven placement contest at the International Symposium on Physical

Table A.5 ISPD-11 (IBM superblue) benchmark circuits

Circuit	#Inputs	#Outputs	#Gates	#Nets
superblue1	8320	13,025	1,098,188	837,712
superblue5	11,661	9617	772,000	754,907
superblue10	10,454	23,663	1,200,000	1,147,401
superblue12	1936	4629	1,995,395	1,520,046
superblue18	3921	7465	898,134	670,323

Design (ISPD), 2011. These are among the largest benchmarks used in SM research discussed in this book. Table A.5 shows some features of 5 of these benchmarks used in SM papers. These benchmarks, also known as the "IBM superblue" circuits, are described by Viswanathan et al. in [15] and can be accessed via [16].

6. Others: Functional designs used as benchmarks in SM research are often drawn from the *opencores.org* website [17], which provides synthesizable Verilog and VHDL models in various functional categories such as processors, crypto cores, communication modules, etc. Another set of benchmarks called the *EPFL suite* [18], which was developed for use in logic synthesis research, is also adapted for demonstrating SM attacks and defense methods.

References

1. F. Brglez, H. Fujiwara, A neutral netlist of 10 combinational benchmark circuits and a target translator in fortran, in *Proceedings of IEEE International Symposium Circuits and Systems (ISCAS 85)* (IEEE Press, Piscataway, 1985), pp. 677–692
2. F. Brglez, P. Pownall, R. Hum, Accelerated ATPG and fault grading via testability analysis, in *Proceedings—IEEE International Symposium on Circuits and Systems* (1985), pp. 695–698
3. D. Bryan, The ISCAS '85 benchmark circuits and netlist format, MCNC, Tech. Rep., 1988
4. M.C. Hansen, H. Yalcin, J.P. Hayes, Unveiling the ISCAS-85 benchmarks: a case study in reverse engineering. IEEE Design Test Comput. **16**(3), 72–80 (1999)
5. ISCAS High-Level Models. https://ddd.fit.cvut.cz/prj/Benchmarks/ISCAS_Desc/ISCAS_HLM.html
6. Collection of Digital Design Benchmarks. https://ddd.fit.cvut.cz/prj/Benchmarks/
7. The Benchmark Archives at CBL. https://people.engr.ncsu.edu/brglez/CBL/benchmarks/index.html
8. F. Brglez, D. Bryan, K. Kozminski, Notes on the ISCAS'89 Benchmark Circuits, Tech. Rep., 1989
9. F. Brglez, D. Bryan, K. Kozminski, Combinational profiles of sequential benchmark circuits, in *Proceedings—IEEE International Symposium on Circuits and Systems*, vol. 3 (1989), pp. 1929–1934
10. S. Yang, *Logic Synthesis and Optimization Benchmarks User Guide Version 3.0* (Microelectronics Center of North Carolina, 1991)
11. ITC'99 Benchmark Documentation. https://ddd.fit.cvut.cz/prj/Benchmarks/ITC99.htm
12. ITC'99 Benchmark Homepage. https://www.cerc.utexas.edu/itc99-benchmarks/bench.html
13. F. Corno, M. Rebaudengo, M.S. Reorda, A set of RT-and gate-level benchmarks, Di Torino, Politecnico, Tech. Rep. http://www.cad.polito.it/

14. F. Corno, M.S. Reorda, G. Squillero, RT-level ITC'99 benchmarks and first ATPG results. IEEE Design Test Comput. **17**(3), 44–53 (2000)
15. N. Viswanathan, C.J. Alpert, C. Sze, Z. Li, G.J. Nam, J.A. Roy, The ISPD-2011 routability-driven placement contest and benchmark suite, in *Proceedings of the International Symposium on Physical Design* (2011), pp. 141–146
16. ISPD 2011 Routability-driven Placement Contest and Benchmark Suite (2011). http://www.ispd.cc/contests/11/ispd2011_contest.html
17. Opencores, Reference community for Free and Open Source IP cores (2020). https://opencores.org/
18. Majority-Inverter Graph (MIG)—LSI—EPFL. https://www.epfl.ch/labs/lsi/page-102566-en-html/mig/

Index

Printed in the United States
by Baker & Taylor Publisher Services